M000012416

The ESSENTIALS

REGISTERED TRADEMARK

# BUSINESS WRITING

**Claudine L. Boros, M.A.**
Adjunct Professor
Business Communications and English
Audrey Cohen College, NYC

Research and Education Association
61 Ethel Road West
Piscataway, New Jersey 08854

THE ESSENTIALS ®
OF BUSINESS WRITING

1998 PRINTING

Copyright © 1996 by Research & Education Association. All rights reserved. No part of this book may be reproduced in any form without permission of the publisher.

Printed in the United States of America

Library of Congress Catalog Card Number 96-69816

International Standard Book Number 0-87891-060-3

ESSENTIALS is a registered trademark of
Research & Education Association, Piscataway, New Jersey 08854

# WHAT "THE ESSENTIALS" WILL DO FOR YOU

This book is a review and study guide. It is comprehensive and it is concise.

It is a handy reference source.

It condenses the vast amount of detail characteristic of the subject matter and summarizes the **essentials** of the field.

It will thus save hours of research and preparation time.

The book provides quick access to the important facts, principles, procedures, and writing techniques in the field.

Materials needed for business writing can be reviewed in summary form – eliminating the need to read and reread many pages of reference materials. The summaries will even tend to bring detail to mind that had been previously read or noted.

This "ESSENTIALS" book has been prepared by an expert in the field, and has been carefully reviewed to assure accuracy and maximum usefulness.

Dr. Max Fogiel
Program Director

# CONTENTS

## Chapter 6
## TYPES OF BUSINESS WRITING: INTERNAL AND EXTERNAL



## Chapter 7
## WRITING STYLE



## Chapter 8
## BUSINESS GRAMMAR

## CHAPTER 1

# Introduction To Business Writing

## 1.1 Business Writing Defined

"Business" writing is the thoughtful writing of letters, memoranda, reports, presentations, proposals, plans, etc., with the purpose of communication for effective business operations and management. The main purposes of business writing is to provide pertinent information, instruction, or guidance for the reader.

## 1.2 Business Writing as a Distinct Discipline

Business writing must meet the same general requirements as all written communication (e.g., clearness, conciseness, etc.).

Writing in itself is an act, just like speaking. Everyone who signs his/her name or jots down a note performs the act of writing. Thus, "writing" is not a discipline in itself.

Few people would dispute that writing by lawyers is "legal writing" (although much of their writing does not qualify as thoughtful legal writing) and is considered to be a separate discipline in itself. Legal writing has its own unique characteristics and language, with its own terminology and writing style (often referred to as "legalese"). Medical writing is also a separate discipline, as is playwriting.

"Legal" is the adjective, the modifier, of the noun "writing"; hence, "Legal writing" refers to certain types of writing in the field of law by lawyers, judges, in specific areas by government officials or legislators, or selectively by authors of legal books and articles.

### 1.2.1 A Comparison between Business Writing

"Business" is an all-encompassing, vast field, and the word "business" as a noun is an all encompassing, generic umbrella term of its many segments, such as accounting, finance, marketing, or production. By sector, "business" includes manufacturing, retail, insurance, and service, among others.

### 1.2.2 Summary Characteristics of Business Writing

The ground floor foundation of business writing is the sufficient knowledge and proficient usage of the elements, principles, and rules of the English language.

Since the consequences of business writing are directly or indirectly monetary, affecting the welfare of organizations and people, it is particularly important that business writing be clear, precise, and relevant.

Business writing is interdisciplinary—e.g., the accountant writes reports to recipients in functions ranging from procurement through marketing.

Business writing requires that the writer sufficiently know the subject matter, the field, and the terminology, as well as the purpose and consequential actions, of the written material.

Successful business writing needs to be viewed in totality. For example, if a particular piece of writing possesses proper English usage, but is weak in terminology or in revealing a clear purpose for action, then it fails at least partially.

## 1.3 Business Writing as a Requirement

Most people will find that even entry-level positions require frequent written communication to other persons within or outside their organization. As they advance in terms of experience, they will experience an exponentially increasing shift in their workload towards the demands of written material issued by them. Moreover, they will find an increasing degree of depth, diversity, and impact from their writing upon operations and management. This will become an important criterion in managerial evaluation of their performance and for their chances of promotion.

In today's widespread world of business, the distance between people has greatly multiplied the frequency of written communication. It has become a common requirement to record almost everything in writing to indisputably document all arrangements and agreements. Indeed, "Put it in writing" is a commonly used business dictum. This has formalized written communications.

## 1.4 Types of Business Writing

The chart on the following page is a selective list of the major types of commonly referenced business writings. Note the absence of items such as invoices, advertisements, bills, statements of accounts, charts, commercials, and publicity announcements. Although these are forms of business communication, this book does not consider them a part of business writing per se.

In fact, most of the types of writing listed in the chart on the following page should be classified as business writings only if they are written for business purposes. For example, financial statements and financial analyses are business writing, even when these are within, for, or about a given non-business entity (e.g., a college), regardless of who issues or prepares them. Conversely, abstracts, agreements, letters, and memoranda ("memos") are business writing only if they are written for business purposes.

## TYPES OF WRITING COMMONLY INCLUDED UNDER THE UMBRELLA TERM OF "BUSINESS WRITING"

Abstracts
Administrative orders
Agenda
Agreements
Announcements
Annual reports

Booklets
Bulletins
Business articles
Business books
Business/financial news

Complaints
Contracts
Cost analyses

Data sheets
Directives
Directories

Educational literature
Employee publications

Financial analyses
Feasibility studies

Handbooks
House organs

Inquiries
Instructions

Job descriptions
Journals

Laws and by-laws
Legal briefs
Letters

Manuals
Marketing surveys
Memoranda
Minutes

Notices

Policy statements
Presentations
Press releases
Procedures
Programs
Proposals
Prospectuses

Questionnaires

Reports
Research studies
Resolutions
Resumés

Sales presentations
Specifications
Speeches
Suggestions

Technical catalogues
Technical papers

In addition, specific organizations have specific needs, and such specialized types of written material may not have been listed here. Indeed, even within the same organization, different departments' written material varies widely (e.g., the written considerations of the financial department are not the same as those of the sales department).

## 1.5 The Surge in Volume of Business Writing and its Related Pressures Upon the Writer

The immense growth of organizations and the catapulting complexities of the business world ("global economy") have resulted in a gradual but steady shift from personal interface to the written communication mode. Most senior executives spend well over 50 percent of their working time reading and evaluating the written business material they receive, writing responses to them, or originating their own business material. The same holds true for many professional staff members. Indeed, much of their workload deals overwhelmingly with writing.

Further, the trend as of late has been to "downsize" company personnel, especially in the administrative and clerical functions. This has placed an added burden upon management to "pick up the slack" in the area of business communication and manage the operations more by written communication than was necessary with fuller staff availability.

The speed available through technological advances (computers, faxes, etc.) has impacted the business world, particularly the business communication arena, more than any other field. Lawyers, literary and non-literary writers, and teachers, for example, have weeks, months, or at least days to plan and prepare their written communications. However, all too often executives, managers, and professional staff have only a day or so to prepare their important written communiqués.

As if the time pressures upon the business writer for rapid action were not enough, there has been a gross overestimation of the value of the computer database for the demands of the business writer's speedy preparation of complex reports, memos, analyses, and many other writings.

All of these factors underpin the great importance for the professionals, managers, and executives in the business world to acquire and constantly enhance their knowledge of the essential elements of business writing. Then they can instantly employ these fundamental techniques for any situational writing, without groping with the task of correct writing methods.

## CHAPTER 2

# The Writing Process as it Applies to the Business Environment

## 2.1 Review of the Writing Process

There are several different arrangements and even outlined descriptions of the steps (flow) of the business writing process. The following is a composite only. The elements discussed as topical in Chapter 2 are expanded upon throughout this book in the context of the various sections, such as Section 6.3, "Business Reports."

Each of these steps are discussed in this chapter in some detail.

STEPS OF THE BUSINESS WRITING PROCESS

I. Planning Stage—Reduce Thoughts into Written Notes

   A. Evaluate the topic of the particular business writing

      1. Decide if business situation is an issue, problem, or plan

      2. Analyze and evaluate data, events, information

      3. Provide information, knowledge, instructions, and orders

4. Other

B. Visualize the audience and their needs

1. List the audience (recipients) for the particular communication

2. Map their education, experience, knowledge, and positions

3. Categorize them by their background and their needs

4. Make other notes about them

C. Brainstorming and strategy

1. Note all thoughts related to the topic; freewriting

2. Conceptualize the strategy for the solution of the problem or developing the information

3. List of ideas, code, or reference words from memory

4. Make other notes

II. Prewriting Stage

A. Organize the Notes and Thoughts From the Planning Stage

1. Map the notes and thoughts derived from planning

2. Review above material, and consider and add additional thoughts

B. Conceptualize and Establish Additional Information Needs

1. Write down what additional information is needed

2. List sources, methods, and approaches to obtain these

3. Separate areas, items, and elements where hypotheses or assumptions will be used

4. Organize these notes systematically, separate known/unknown

5. Begin to obtain the information

6. Begin to formulate the hypothesis, assumptions, and premises

C. Prepare Outline for the Communication

1. Type and nature of correspondence governs: report, memorandum, proposal
2. Prepare preliminary (from scratch) outline, table of contents, or other sequential listing
3. List the appendix material planned as documentation or illustration

III. Drafting Stage

A. Write the Preliminary Draft

1. Review the notes and preliminary outline from the planning and prewriting stage
2. Prepare a "core draft" as an expansion of the outline
3. Write the first draft: freewriting
4. Obtain the needed information parallel with writing the first draft

B. Discover and Adjust Preliminary Draft

1. Adjust the first draft for new ideas and new information
2. Check the hypothesis and coherence of the first draft
3. Review and adjust the first draft for purpose and audience

IV. Revising Stage

A. Revise the Draft for Content

1. Reread the first draft several times
2. Revise the content for accuracy and completeness
3. Revise the content for purpose, message, and audience needs

4. Prepare or revise the introduction, conclusion, and recommendation

B. Revise the First Draft for Writing

1. Revise the draft for unity: coherence, transition

2. Revise the draft for all aspects of English

3. Revise the draft for clarity

4. Revise the draft for tone, decorum, and style

V. Edit the Revised Draft

VI. Proofread the Final Draft

VII. Issue the Final Correspondence

## 2.2 Time Management

The amount of time available and the deadline vary greatly for the business writer, dependent upon each written communication. While many memoranda and reports need to be prepared almost instantaneously, many other communications (reports, plans, proposals, analyses, and letters have a sufficient timeframe for the business writer to go step-by-step through the writing process when preparing the document.

The business writer should plan the timeframe of the work required for each written communication. He/she should prepare a scratch "time-plan" as to what part of the writing process is targeted for completion by periodic days or dates, and try to adhere to or even beat these self-imposed deadlines. It is also advised that the writer allow additional time to compensate for any problems that may arise.

The setting of the timetable itself is a circuitous process, from the Planning Stage through the Final Copy.

## 2.3 Planning and Prewriting Stage

It is important for the business writer to complete the Planning and Prewriting Process prior to formulating or composing his/her conclusion (thesis) presented in the written communication. An effective conclusion should be derived from the content of the memorandum, report, or letter; it should make a compelling statement about the content itself. This is expounded upon in Chapter 3.

If the writer first composes the conclusion, then the danger becomes that he or she will write to its specifications and manage the text to fit the advocated conclusion, often resulting in distortion of fact, analysis, interpretation, or advice.

Some of the recipients of such slanted communication will be knowledgeable and experienced executives, managers, or professionals, who will detect that the report has been stage-managed by the writer. Such members of management often conclude that even if most of the report is factual, accurate, well-analyzed, competently interpreted and logical, they still cannot accept it on the whole because of the fallacies incorporated in the text.

The business writer should bear in mind the above advice, because it is all too common—and not only in business—to write from the writer's mind-set and emphasis, relaying the picture as the writer sees it. Objectivity in business writing is a cardinal rule. Misleading the recipients into believing that what they read is factual information, an objective presentation, or an unbiased opinion or recommendation is unprofessional and harmful to the business writer and his/her department or company.

Here is some essential advice for the business writer:

- In each business situation about which the writer is to write, the first event is the thought process that takes place in the mind of the writer. Since many of these thoughts hit the writer like flashes of lighting, he/she is able to capture them only by jotting them down or dictating them on tape as they occur, even if only in a fragmentary manner by listing code words, dates, and numbers.

- The same applies to brainstorming. The writer should jot down whatever comes to mind throughout the development of the information, drafts, and revisions. Freewriting in the Planning and Prewriting Stage means just what its name implies—the writer should not concentrate on grammar, good English, or coherence. The writer should write from memory the known information and other points in unrelated segments.

## 2.4 Drafting the Written Communication

There are basically two approaches to drafting the report (here we use "report" selectively for "written communication" for easier connection with the reader):

1. Begin with and proceed to write the first draft in strict sequence and adherence to a pre-formulated and well-established outline or table of contents. (By the Drafting Stage, these must have been revised several times into a near final version.)

   Using this approach, the business writer begins with and completes say I. A. 1., then I. A. 2., and so on before proceeding to the next point in the sequence. Even here, however, he/she should not get tied-up with English language aspects such as diction, grammar, or mechanics.

   This approach is hindered if new information is still being gathered during the Drafting Stage, which might unravel parts of the draft itself.

   Here, the draft will be a rather complete report, ready for the Revising and subsequent Editing stages.

2. The other, more effective, approach is to write the first draft rapidly and continuously, with the idea that it will be adjusted into a second and third draft before the Revising Stage. This is like an investigative or discovery drafting, done in piecemeal, not in sequence. Those parts that stand in the forefront of the writer's mind are written first. (Even here, some authors advise the writer to first write the draft for the most difficult sections of the text.)

After completing the rough first draft, the writer should review and rewrite the draft, checking it against his/her "checklist"—a task facilitated by the computer/word processor.

One of the advantages of this approach to the draft is that when additional information is being obtained parallel to the writing of the report, or when new ideas, concepts, or solutions emerge during the writing, then these can be readily incorporated into the draft, without resulting in loss of time and effort.

In the second and third draft, the report should emerge meeting all the standards that are set forth throughout this book for professionally written communication.

Foremost, the text must be correct, accurate, objective, and logical, with supportable conclusions and (where applicable) prudent recommendations. The objective of business writing is not to write in good English per se, but to provide information, instructions, solutions, and recommendations that benefit the business (company, organization). Competent English, however, is essential for clarity, readability, and correctness.

## 2.5 Revising the Draft

Revision of the draft is a process different from and proceeding the stage of editing.

Basically, revising any draft requires that the business writer revisit from scratch each part of the report, as if he/she is its auditor (not to be confused with editor). The writer must divorce himself/herself from the writing aspect and scrutinize the text for the validity of its premises and propositions; content, facts, information, and opinions (i.e., the evidence); descriptions, explanations, and arguments; and conclusions and recommendations.

The best approach to this revision process is to view the report through the eyes of the future recipients, their expert staff, and outside "experts" (if any). This is a critical review by the writer.

Frequently, in large companies, there is an informal or formal system whereby at the Revising Stage, one or more other business professionals review and advise the writer of necessary revisions.

The final product of the revision process is the final draft.

## 2.6 Editing the Final Draft

The first purpose for editing by the business writer is to ensure conciseness of the text. Not brevity—but conciseness—for the two are different, as is discussed in this book. Brevity is ill-advised in business writing.

Then, the writer is ready to edit the text for all aspects of English: grammar, diction, vocabulary, punctuation, and mechanics. However, business writing is not common English writing, for business has its own professional language, not dissimilar to that used in literature.

Yet, in most books on business writing, authors include lists of words which they usually call "jargon," and advise the businessperson to substitute common and popular English words for them. For example, the words "execute," "facilitate," and "terminate" are far closer to common (popular) English than is the term "personification" or the truly abstract term "symbolism."

Thus, editing should be approached most carefully and from the vantage point of business professionals, taking into account their views and language.

## 2.7 Proofreading

Many reports, presentations, or other forms of business writing are reproduced and disseminated to numerous recipients. Procedures and handbooks in a large international or decentralized company may end up being distributed to well over 100 recipients, many of whom may use English as a second language.

The business writer should ensure in the proofreading stage of the writing process that the text is error-free, particularly with regard to the text's meaning and completeness.

Business writing books and even business communication college-level courses seldom address the point that the business writer, thought greatly aided by the use of advanced computers and word processors, often gains a false sense of security from such devices and thus tends to neglect the proofreading process. Detrimental errors are therefore not always purged from the text.

Proofreading should include:

- Checking to ensure that words, lines, paragraphs, or even pages were not omitted or left incorrectly in the final text. Using the computer (or word processor), writers move around portions of the text, revise the text, and edit the text while it is on the computer screen. The writer needs to proofread his/her own text, even after the secretary or staff member has proofread it.
- Making full use of the computer's spell-check program to catch misspelled words. However, there are many words that are not included even in the most advanced spell-check programs. These include names (of specific companies, individuals, products, etc.), numbers (including I.D. numbers and codes), technical terminology, and foreign words.

  Moreover, this is where a particular pitfall of proofreading lies. For example, the simple words "to," "too," or "two" are each spelled correctly, and the spell-check will not alert the writer if one of these words is used in place of another. But the wrong word—especially if the text is dictated—may lie undetected in the text. Again, burdensome as it is, only the writer can ensure that such errors are not present by proofreading the final text.

## 2.8 Final Copy

After the draft is subjected by the business writer (and perhaps others) to the processes of revision, editing, and proofreading, the draft becomes the final draft.

The final draft must be then checked or arranged (if it hasn't been yet) into the desired format, and adjusted for all the ingredients of the business letter, memorandum, or other type of communication. These encompass the typed format (such as full-bloc) and margins to the arrangement of lines ranging from "dateline" to "enclosures (encs.)." These aspects deal primarily with the appearance, the "look," of the written material which is discussed particularly in Chapter 6 as to its importance and arrangement. (The business writer does not need to be diverted here with a discussion of margins, type of paper, format of "full-bloc," or page numbering; these are routine matters that are handled by typists and secretarial staff.)

The final draft is then printed as the final copy and distributed to the recipients. It may sound pedestrian, but the business writer should ensure that each recipient receives the written material a complete copy of when it is sent to several individuals. Burdensome as it may be, after all the time and effort invested by the writer in preparing the written communication, it is only prudent for him/her to ensure that each recipient receives the full package. (It may sound like a clerical point, but high-speed copiers often skip or tilt pages while running them off, and incomplete or flawed sets have been mailed, to the frustration of recipient and writer.)

CHAPTER 3

# Purpose of Business Writing

## 3.1 Principles of Business Writing

The business writer must ensure that the recipient of the written communication comprehends the message that is intended by the writer. The cornerstone of this is that the writer clearly conveys the message that he/she intends. The writer should not merely try to demonstrate his/her brilliant literary style—business writing is not writing for writing's sake.

This book emphasizes concise (edited), focused, stream-lined, correct, and unslanted, "professional" writing by the business writer. However, conciseness must not be confused with brevity—the writer should not condense correspondence, aiming for one-page documents as the ultimate goal. While text should be edited for length, it still must sufficiently include everything that is relevant and necessary for the recipient's understanding for action.

### 3.1.1 The Image: The Written Message Is for the Recipient's Benefit

It is important for the business writer to plant the image in the minds of the recipients that they, or their department, group, or business, will benefit if they act in accordance with the writer's advice. The writer should force into the background and camouflage the image of his/her own benefit from the recipient's action which the writer desires to provoke. The recipient's viewpoint and benefits not the writer's should be reflected as important. (The opposite is true, however, in the case of directive, instruction, command-type letters, memos, procedures, handbooks, or manuals wherein the writer's own viewpoints, or those on whose behalf he/she writes, should be reflected as directives.)

### 3.1.2 The Writer's Image of the Recipient

The writer should construct a mental image of the recipient of the written communication. In the Prewriting Stage (see Chapter 2) of the letter, report, or memo, the writer should make notes of the main characteristics, data, or information known about the recipient and, if necessary, find out more about him/her. Psychological factors (known or conceptualized) need to also be considered in many cases.

Generally, the business writer knows considerably more of these factors (or has the information available to him/her) when the memo or report is addressed to recipients within the writer's own company or organization ("internal"), albeit this advantage is certainly reduced when the writer is employed in a large, multilocational (especially international) company.

In such circumstances, these variable factors are multiplied when "writing across borders" and/or writing to interdisciplinary recipients (for example, if the Cost Accountant writes to General Managers, Controllers, Marketing/Procurement/Legal/Manufacturing discipline recipients). This factor is discussed in detail in Chapter 5.

### 3.1.3 Purpose of Writing

Business correspondence has one of two basic purposes, either short-run or long-run. The short-run purpose business letter or report is aimed at inducing the recipient to act upon whatever the business writer proposes, requests, or suggests the recipient do, be that a positive or negative act. The Long-Run Purpose business letter or report also has the aim of action by the recipient but at a delayed time; its aim may also be to provide information, advice, planning data, etc.

### 3.1.4 Write Only When the Situation Warrants It

Many letters and reports have been rewarding to their business writers in a great spectrum of circumstances and at a wide range of levels. Unfortunately, these "success stories" have resulted all too often in persons writing letters and reports which—at least at the early stage of events or due to lack of sufficient information—should not yet have been written, because they resulted in partial or total failure for the business writer in his/her communication or intent. Many written communications should never have been written at all, because they were not warranted. Conversely, letters or reports should not be deemed by any businessperson as a burden or chore, but viewed as opportunities to communicate his/her message.

Corporate policy manuals and procedures often govern what the business writer documents and when he/she writes it. While such policies and procedures provide valuable guidance, the business writer must understand when and where he/she might deviate from them, as necessitated by the specific business situation. If necessary, the writer must obtain approval from authoritative managers or executives to deviate from the corporate policy manuals and procedures.

### 3.1.5 Primary Importance: The Writer Must Know What He/She Is Talking About

The writer must verify, check, "audit," and scrutinize all details in the written communication and state as factual or accurate only those which are factual or accurate. If the writer does not have the ability to ensure that the information is absolutely accurate according to factual or authoritative bases or grounds, then he/she must qualify this in the text, in order to enable the recipients to understand what reliability percentage they can attach to amounts, dates, concepts, facts, or other elements.

Nothing can be more detrimental to the specific business situation—and to the business writer—than inaccurate, assumed, unreliable, unconfirmed, or incorrect data projected by the writer to the recipients as if such were factual and accurate.

“That person knows what he (or she) is talking about” is a compliment that the business writers should strive to have said about them. Such a compliment is reservedly used in the business world to describe writers of reports or memoranda, even in cases where it is expected to apply.

### 3.1.6 The Other Side in Communication: Comprehension by the Recipient

It is stated in this book that the critical factor in business writing is what the recipients grasp, comprehend, deduce, or otherwise attain from the written material. Readability is an essential goal of business writing.

Readability in business communication means that the content of the written material is clear to the recipient. This aspect is essential to business writing. It is presented in this book that the audience, i.e., the recipients, of the business writing (be that letter, report, other) vary extensively. In addition, global competition and company downsizing have resulted in time pressures upon almost everyone in business, eliminating the more relaxed atmosphere of a mere 30 years ago.

Recipients of weighty or lengthy business reports/memoranda should be able to clearly and sufficiently understand them, even if they don’t know anything about the technical subject matter present in the writing.

## 3.2 The Situation Governs the Writing

Recently, General Motors Corporation (GM), navigated by its highly educated in-house and outside lawyers, CPAs, governmental affairs consultants, and an array of other extensively experienced businesspersons, planned to establish a large manufacturing operation in Poland, under complex conditions which these professionals had developed.

A covenant of this planned investment was that the Polish Government agreed to waive its 35 percent customs duties on GM products. This 35 percent customs duty rate was assessed on the products of other automobile companies (except for Fiat and Volkswagen, which were to obtain a similar deal from Poland).

However, this extensively developed arrangement failed for GM because the company did not learn up front that this practice was contrary to the GATT agreements. GATT headquarters ordered it to be canceled, and GM had to rearrange its plans. (For a laymen's understanding as to why it was contrary to GATT, it was discriminatory against the other automobile companies.)

What happened was clearly that this *business situation* was unique to itself, and everyone involved at GM lacked specific detailed knowledge or did not consider the GATT agreement and clauses relevant to this specific business situation. These individuals did not consider this in their scope of review in the Prewriting Process.

This is not an isolated occurrence. Many other cases involving different parties and circumstances could have been selected to illustrate this point, but the above is a telling example since probably at least a hundred financial, legal, operational, and other—including these companies' in-house and Poland's customs professionals—experienced persons were business writers and/or recipients of the written business communications that dealt with the above situations.

The lesson for business writers is that when they are required to write about a specific business situation, they must adopt the mind-set in the Prewriting Stage that they need to identify all factors, points, and elements which *might be* involved with the project. The business writer must then approach each of these factors, points, and elements as if such factors are unknown to them, adopting a sort of "writing from scratch" approach.

In summary, the business writer must not only draw upon his/her knowledge but must comprehend each business situation through inquiry, investigation, and identification of causes and effects involved within the framework of the objectives of the project.

## 3.3 Inform, Report, Explain, or Advise

Forms of business writing where the writer informs, reports, explains, or advises the recipients on business situations, subject matter, or facts are informational, both in terms of their nature and with regard to their objective. Some books on the subject categorize such writing under the term "informative writing."

### 3.3.1 Inform

Informative writing is not restricted to any one type (medium) of business writing. It may be found in a letter, memorandum, or report. While it is proper to refer to these as "informative writing," it is incorrect to think of them as providing information only.

The essential characteristics of informative writing are:

- *Objectivity.* The business writer should present the information without introducing his/her own judgments, recommendations, or viewpoints about the subject matter.
- *Factuality.* Only factual information should be presented, be that about events, concepts, methods, or any other subject matter about which the writer informs the recipients—they should evaluate and draw their own conclusions or act upon the information received.

Often, however, when the business writer prepares the letter, memorandum, or report with the purpose of providing information, by the nature of the subject matter (or by the writer's explanation of it) the result is that this writing becomes instructive or even persuasive.

A helpful distinction of *informative writing* is that it *is not intended to be affective writing;* that is, writing in which the business writer attempts to influence the recipient in some manner.

### 3.3.2 Report

"Report" basically means that the business writer conveys information, "reports it," to the recipient in an orderly and objective manner, with or without his/her analyses, recommendations, or conclusions. The report writer's main task is to present—to relay—factual information.

### 3.3.3 Explain

Frequently, the task of the business writer is not only to provide information about events, facts, situations, concepts, or other matters, but to go beyond this and *explain* to the recipients what this means to them. This is defined as explanatory writing.

Often, unlike in informative writing, an executive or other member of the company is the one who turns over the matter to the business writer for his/her explanation to them; thus, the information does not even originate with the business writer.

For example, as the outcome of a certain meeting, the attendees request the business writer (a planning analyst) to explain in writing what zero-base budgeting means (say someone talked about it at the meeting). Another example: the business writer (an in-house attorney) reads a newspaper article which reports the dumping exposures of a competitor company. He/she writes an explanatory memorandum to several executives, managers, and staff members whom need to understand such dumping matters.

The information must be factual and accurate, and the business writer must provide an objective, "expert" explanation—not one biased by his/her feeling (like in the dumping example) nor one expressing favoritism or antagonism (like in the zero-base budgeting example).

### 3.3.4 Advise

The most comprehensive form of informative writing is when the business writer:

- Provides information in a memorandum or report to the recipients;
- Explains the information to them; and
- Advises them as to what it means to the company itself—to its "business."

The term “advisory writing” is this author’s own term for such comprehensive letters, memos, or even, at times, reports (not to be confused with the “formal report” discussed in Section 6.4).

Advice is frequently provided to others in letters or memoranda by a wide variety of business persons. Advice is not a recommendation, nor should it be written in a persuasive, inducing, or affecting tone, or with such intent.

For example, in the above dumping case, in addition to providing the information about the competitor company’s exposure to extraordinary assessments by the U.S. Government (i.e., dumping charges), the writer explains the grounds and reasons for dumping assessments. The business writer then advises certain members of his/her own company (and perhaps also the outside law firm of the company) why there may or may not be a similar dumping exposure to his/her company. To wit, the business writer should provide only his/her advice as to such applicability and omit expansion into recommendation for changes in the company’s business transactions or even discussion of the philosophical issue of the validity of dumping.

Likewise, in the case of the “zero-base budgeting,” the business writer should provide only his/her advice as to how it can apply to the company, but omit his/her preference or opposition to it, if it is to be considered advisory writing in purpose.

## 3.4 Instruct, Direct, or Order

The business writer at times must write instructions, directives, or even orders (mandates) to others, to do or refrain from doing certain things. The language in such writing needs to be commanding, although in tone it can range from being instructive to authoritatively commanding, dependent upon the nature of the written communication.

It is noteworthy and relevant to set as a premise that the vast majority of business professionals who are the “business writers” of the managerial level of instructions, directives, and policy orders are only the writers. Such written business communications were delegated down to them by their superiors or by committees or management groups, who selected them because of their technical professionalism in the subject matter. Most often, such written communications are released under the signature and authority of a higher executive than the particular business writer. This makes most business writers of such essential communications "ghost writers."

In such writing, the business writer must not use language that indicates or seeks cooperative agreement from the recipients. These are not suggestions by the writer, but instructions, directives, or even downright orders, and it is not optional to the recipients to choose to agree or disagree with them. (At least, that is the writer's *purpose*, although he or she might not be involved in the process of ensuring compliance with the instructions.)

Persuasive and personal tone must be absent from this type of writing. These are not intended to be courteous or complimentary forms of communication.

The tone of the language in procedures, directives, and policy orders is *not* of the "please," "if you would," or "we would appreciate it if you..." variety. In procedures and directives, usually the verb *will* is used, which all recipients know, or soon learn, means "will adhere or face the consequences" (whatever they may be). *Must* is a word often used in lieu of *will.* At the highest strata, such as policy orders that cover laws, regulations, or the chairman's or the board of directors' mandates, the command verb *shall* is often used, in order to elevate these orders to a plateau even above the mandatory instructions.

The business writer *must not* deflate the tone of these command words by using instead the more polite and palatable "would, should, could, can, may"—these are escape clauses for the recipients as if it is left up to their discretion to oblige, perform, or adhere to the instructions, directives, or policy orders.

### 3.4.1 Instruct

When writing instructions, it is essential that they are written in the proper sequence, step-by-step. The business writer must assume the place of the recipients and critically evaluate if they would clearly understand the totality and each element and step of the instruction.

If the writer does not have a sufficient knowledge or understanding of any aspect, element, or step involved with the instruction, then he/she must ascertain from others what the proper instruction needs to be.

Many instructions have been issued in business that had to be altered, revised, or even revoked, because they could not or should not have been implemented or performed, despite the fact that they were favored or desired at the time they were issued. This is an adverse reflection upon the knowledge of the business by the business writer and his/her department.

Procedures are clear examples of instructional writing, because they require step-by-step instruction in an orderly and sequential manner. For example, in a procedure on Purchasing Department sourcing, the first step probably is the instruction for Purchasing to find several suppliers who produce the wanted material or product. It would be incorrect to set as the first step the price criteria consideration.

The business writer should also foresee and include in the instructions, especially in procedures, the guidance for flexible performance under circumstances differing from those set forth in the procedure or instruction. A businesslike approach for the business writer is to include in a section, preferably in the introduction, that should conditions be different, then the recipients must contact the controlling authority to seek instruction under the changed situation.

Less frequently, the instructions or procedures are issued for something that the recipients "will (must) not do." For example, not to use the division's funds to speculate in the foreign exchange market, which is the corporate treasurer's prerogative.

### 3.4.2 Direct

In instructions—such as procedures—the common element is that the business writer provides step-by-step descriptions of what must be done or what must not be done.

Directives usually do not involve such step-by-step procedural instructions for performance. Directives are usually company mandates for matters that do not cover governmental laws or regulations but are under company policies. The business writer should use the "must" or "shall" command words, or at least the "you *will*" command, to make directives in instructions clear. The following examples illustrate this:

- The Corporate Financial Vice President and Chief Financial Officer (CFO) issues a directive to the general managers of the foreign subsidiaries of the corporation, whereby all of them must (will or shall) submit to the CFO's office for approval any planned or desired price deviation from the company's published price lists.

  The business writer (in this example the pricing policy Manager) writes the directive on behalf of the CFO. This manager needs to set forth in the directive what is meant by the price lists, and under whose authority these have been issued in the corporation. The manager should add in the directive that the general managers also must submit to the CFO's office for approval any prices for products that somehow do not appear on the price list (a new product, a by-product, which emerges from production of a regular product).

### 3.4.3 Policy Orders

The strongest tone and language in business communication is used for policy orders (these have several different titles among companies), which are mandates by corporate management for the employees (including executives) of the company. Here, the business writer should use the commanding, imperative "shall" to elevate these communications on the highest plateau.

Most often, such policy orders are issued in conjunction with laws or government regulations that are in the forefront of the public eye or in the business world at a given time. For example, in the 1970s, bribery of foreigners by American (and foreign) business-persons or their agents or representatives emerged as a public issue, resulting in the U.S. Congress' Foreign Corrupt Practices Act. In every internationally involved company, a policy order was issued under the Foreign Corrupt Practices Act, forbidding such practices to everyone working at or associated with the company. These were invariably issued under the name of the Chairman or President and CEO, but written by whoever was the business (ghost) writer.

The business writer of such policy orders needs to write specifically what is prohibited by the company (in this instance ranging from entertainment to cash bribery), and to cite and describe the grounds for this prohibition (in this instance the Foreign Corrupt Practices Act). This is not "informative writing," but it does have similar aspects in the essentials of providing accurate, correct, and documented information to the readers, as the support for the policy order.

### 3.4.4 Guide

One of the most frequent types of writing in business is memoranda or mere informal notes in which a person knowledgeably provides guidance to others about actions, proceedings, performance, reporting, or other matters.

Unlike instructions, directives, or orders which are the prerogatives of a select few specialists, guidance is written in business by persons from a wide range of levels and positions.

Most often, guidance is technical in nature as to how to go about carrying out various tasks from Step 1 through the end process of the subject matter. (As throughout this book, under "technical" we do not restrict it to engineering or scientific fields—the controller gives accounting technical guidance to accountants, for example.) Thus, guidance writing should be sufficiently detailed—down to the nitty gritty, as it were—to provide informal instructions to the recipients to follow step by step. It is similar to in-house teaching by instruction.

It is a useful technique in guidance writing to map out everything involved with the subject matter, and then select what should be included in the written guidance and how it should be presented. This enables the writer to take into consideration as many factors as may be relevant. For example, in "mapping-out" for guidance to the purchasing managers and staff the method of purchasing timely the components, materials, and supplies, the writer should discipline himself/herself to consider the contingencies, such as supply line interruptions by strikes, weather, or production breakdown, which extraordinary factors he/she might not consider unless going through the "mapping-out" phase.

The language of guidance writing should be professional, particularly precise, and utilize an impersonal tone, as if giving exact travel guidance to a person to go from point to point. The recipients are to follow the guidance because of the hands-on expertise revealed in it, although often they had better follow it because it is written by or under the name of their superiors or persons of professional authority in the company (i.e., a staff accountant, a quality control specialist, etc.).

## 3.5 Persuade, or Induce

Viewed in the broadest sense, all written business communications are persuasive. Even in cases where the business writer only provides information, he/she persuades the recipient to acquire new knowledge.

It is common in books on business communication and business writing to present persuasive writing mainly for the purpose of sales, advertising, publicity, promotion, employment, and settling credit or collection disputes. This is persuasive writing to influence the attitudes, behavior, beliefs, or habits of those receiving the letters, brochures, bills, employment applications, or similar personal-oriented communication in favor of the writer's company. This is the "selling" of products and services by persuasion.

The central focus in such persuasive writing is on human behavior. The motives, feelings, and often social conscience of the recipients are primarily considered, and the written communications are prepared to target these behavioral aspects. The language of such writing gives consideration to these factors, and projects recognition and appreciation of the recipients for their own reading and consideration of the letter, advertisement, or application.

Even in the case of bills (invoices), the aim is to show why these should be paid (such as stamping on the envelope "Past Due" or "Have you overlooked?") for reasons of social conscience.

Such writing, however, is not in the realm of business communication, and does not meet the requirements of "business situations" (except if "sales" involves "marketing").

### 3.5.1 Persuasion in Business Communication

The fundamental element of persuasion in business communication is for the business writer to visualize and determine the link between the proposition at issue and the mind-set of the recipient towards the subject issue. This becomes more complicated when the written communication is sent to several individuals.

Particularly in interorganizational (internal) business communications, the business writer basically has to exclude appeals to personal feelings, personal needs, emotions, attitudes, and social motives. The recipients receive the business communication because of the positions they hold; they are expected to act objectively, for the good of the organization, not on whims of human behavior. Nonetheless, the business writer should consider factors of ego, personal goals (that may not always be best for the company), and even feelings (including "dual loyalties," such as towards company/employees, company/community, or country).

The essential rules of persuasion in business communication are discussed here in summary only. The reader needs to visualize business situations for his/her further contemplation about persuasive writing, particularly by recognizing that these are set apart form the "selling" type of persuasive letters, brochures, and advertisements.

The business writer is in the fortunate position that he/she does not have to be creative to get the *attention or interest* of the recipient to read the memorandum, report, or other communication. Internal recipients are a "captive audience"; they must read what they receive from the writer for the common interest of the company.

Even "external" recipients are expected to read the written communication—or delegate it to others to read—as it relates to or affects their own company or organization (governmental or private).

The *content* of the written communication is (or should be) the "attention grabber." Content is further discussed in this section with regard to its detailed elements. However, the business writer must carefully select the recipients based on their need to be informed of the "content." (This is discussed throughout this book.)

Depending upon the specific nature of the particular business communication, the *motivation* of the recipients include: 1) solutions to problems, 2) satisfaction of business needs, 3) increased business knowledge, 4) new opportunities, or 5) other aspects (to be identified by the business writer).

The business writer needs to visualize and determine the appropriate and relevant *motivational factors* in each situation—*not* on his/her end, but as they relate to each recipient (and primarily to the decision-making recipient).

This is made more difficult for the business writer by the fact that often the negative options or even opposition of the selected recipients are known to the writer, but it is *prima facie* evident that these are not their real motives, and that there are "hidden" motives. *Example:*

- Corporate management decides to have an outside group conduct global business communication seminars at each division, with the attendees to be specified level position holders, including the general managers. The Human Resources Vice President is assigned to write the memorandum to establish the program and logistics at the divisions. He/she learns about the opposition of several general managers, who give the reason that it would disrupt operations.

  However, the business writer here is a savvy and experienced manager in human services and personally knows each general manager. He/she knows that they are opposed to the seminar, and that their "hidden" reason is that they staunchly believe that corporate wants to use the seminars for the true purpose of testing, evaluating, and grading the divisional managers and staff for language, writing, and other skills. That is, corporate has its own "hidden" agenda.

The Vice President in this business situation needs to persuade the general managers that their fears are unfounded, and that they should welcome and support the seminars at their own divisions. The connecting "link" between the proposition (actually a corporate mandate) and the change in attitude by the general managers can be the Vice President's convincing point that down the road the Chairman's office will require all divisions to report and otherwise communicate within the company in the precise manner presented by the outside group at the seminar. Thus, they will be prepared to report and communicate properly, avoiding problems in the future with the Chairman's office.

This is a rather simple example of a business situation where *motivation in persuasion* is a key element. A much more complex example would be for the Corporate Marketing Vice President to persuade, the company's Japanese subsidiary General Manager to phase out production of Product XYZ and import these products from, say, the French subsidiary. Assume that the Marketing Vice President projects that it would be beneficial for the Japanese subsidiary to change from making the products to importing them. The Japanese General Manager and staff would naturally oppose such a switch.

These two business situations are vastly different, and so are the respective persuasive writings, including the directions the two vice presidents need to take if they want to persuade and obtain agreement. This will be made clear in the following subsections.

Sufficiently thorough *narration, description, and explanation* of the topic or issue (system, plan, method, product, etc.) is often, but not always, essential in persuasive writing. It stimulates the recipient's interest to benefit from the proposition. Therefore, the writing must be sufficiently thorough and correct, as well as be written in clear English.

Obviously, in the case of the seminar example, the description and explanation is essential to persuade the general managers that the outside group will expand, elevate, and enhance the professionalism of the divisions' business communication capabilities. This requires the Human Resources Vice President to go through the writing process in order to develop the memorandum to the general managers. He or she has to demonstrate in the description and explanation how and why they need the seminar, despite their on-going, already extensive, experience with business communications.

Conversely, narration, description, and explanation is almost perfunctory only by the Corporate Marketing Vice President. The Japanese subsidiary General Manager knows as much and in several areas much more than this Vice President knows or can tell the General Manager about the product and the operations.

The essentials of persuasive writing in business communications are the *facts, options, and evidence.*

The facts must be verifiably accurate, reliable, clearly stated, and, above all, relevant to the problem or issue involved in the specific business communication.

The evidence is fact underpinned by expert opinions.

Opinions must be from persons considered to be experts on the specific problem or issue, and opinions must be unbiased.

At times, the evidence is made more by the relevant experience of other companies (that is, relevant to the problem or issue involved with the business communication).

In the example involving the Corporate Marketing Vice President as the business writer: he/she should persuade and convince by including facts and expert opinions, and providing credible relevant experience of other companies to demonstrate to the Japanese subsidiary General Manager that the orderly discontinuation of the production of Product XYZ, and the substitution of imports from the French subsidiary is in the best interests of both the Japanese subsidiary and the company.

"The facts speak for themselves" is a common saying, particularly in the world of business. But it is often a false saying.

The facts, opinions, experiences, and evidence drawn from a particular piece of written communication can result in different conclusions reached by various persons, if they are left on their own to do so. Therefore, the burden of proof is on those who assert, claim, or propose. They must prove their assertions by *argumentation.*

The skillful business writer presents his/her arguments in favor of his/her proposition throughout the text, instead of pushing them all to the end of the conclusion section. As discussed in more detail in Sections 2.3 and 6.3, the writer should use *inductive or deductive reasons or arrangement* to lead the recipients to the writer's desired conclusion.

Since persuasive writing in business is the most widespread *purpose communication,* it is important to repeat here the principles of these two approaches to *logical reasoning.*

- In both arrangements, the test or standard is the *validity* of facts, opinions, and examples. But the writer must prove the validity of the explanations and arguments that he/she proposes, advocates, asserts, or claims as the proper *conclusion and action.*
- In the inductive reasoning (arrangement) approach, the writer leads the recipients to reach one or more general conclusions from the examination of facts, as aided by the explanations and arguments. The conclusions should encompass (embrace) only what the evidence supports or warrants. The same is true for deductive reasoning.

  In the instance of the Japanese business situation example referenced above, the Corporate Marketing Vice President should use the inductive arrangement to present several facts, opinions, and examples as evidence. This is to be associated with his/her arguments and logical reasoning as to why it should be concluded that the action of discontinuing production and importing the same products from France is best for the Japanese subsidiary and/or for the corporation overall.

- In the deductive reasoning (arrangement) approach, the writer provides up front his/her generalized conclusions and then guides the recipient to reach the same conclusions by the end of the text.

  In the instance of the seminar example, the Human Resources Vice President should use the deductive arrangement, by stating his/her conclusion in the introduction that business communication seminars conducted by reputable experts are deemed important by many companies.

  This conclusion is then supported in the text of the memoranda to the general managers through facts, opinions, and examples as evidence, in which the writer argues by logical reasoning that all general managers have to conclude that the seminar is needed by their divisions as well. That is, it is established that such seminars are generally beneficial to companies, and it is argued and demonstrated in the memorandum that the same applies for each of the company's divisions.

The objective of aim of persuasion is *action* by the recipient in line with what the business writer sought to effect. The writer can enhance this by clearly identifying what the action should be, and by providing in the text helpful guidance of how to carry out the action. (Such guidance varies by instance, ranging from general to specific, and from limited to extensive.)

It is of critical importance that the business writer persuade the recipients through the process described in Section 3.5.1, by adhering to the principles of: 1) objectivity, instead of being overwhelmed by his/her personal stake or bias in the outcome; the writer must even guard against the recipients' perception of such bias on the part of the writer; 2) providing the recipients with all pertinent and known (to the writer) facts, opinions, and examples, including those which are unfavorable towards the writer's proposition; and 3) giving correct, accurate, verifiable, and sufficiently complete information as evidence.

Regarding point 1, it is prudent for every business writer to maintain a certain distance from the subject matter. This acts as a sort of buffer between the writer and the situation in case the subject proposition, when implemented, isn't as successful as foreseen initially at the time of the written communication. "Overselling" by powerful persuasion of plans, ideas, systems, or whatever is at issue often comes back to haunt the business writer, because the recipients can point to him or her as the catalyst for persuading them to act.

Regarding point 2, aside from not knowingly withholding any information (i.e., which is known to the writer at the time), it is prudent for the writer to denote partial information and areas where further information should be obtained.

Regarding point 3, it is essential that the business writer present as facts only those for which he or she had documented evidence or support, and were verifiable and correct upon the writer's own scrutiny or investigation. Concerning information of lesser strength, the writer should include qualifying comments as a caution to the recipients.

### 3.5.2 Induce

The word induce is generally thought to be synonymous with the words persuade and convince. The word induce, in fact, applies to situations where the business writer provides an inducement for the recipient *to act* according to the proposition of the writer.

For example, in the case of the seminars if the Human Resources Vice President so wished (or was instructed by the Chairperson), he did not need to persuade the general managers in lengthy exposition about the values and merits of the seminars. This Vice President (the business writer) could have simply *induced* the general managers to have the seminars by stating that the Chairperson would directly reprimand those general managers who did not conduct the seminars at their divisions or if the future corporate standards of business communications were not met at these divisions.

# CHAPTER 4

# The Language of Business and the Writer

## 4.1 Business Terminology and Usage

While it is valid to emphasize the importance of competent English usage, the business writer is judged by the professionalism depicted in his/her letter, memorandum, report presentation, manual, or handbook.

In fact, such professionalism navigates the writer into a writing style that has been criticized all too often in books as one that desires to obscure ideas by using pretentious lingo to show off his/her knowledge of language to the uninitiated. Much of business, technical, legal, and governmental writing is characterized by such language and terminology—"jargon" as per the dictionary definition. However, such criticism is totally unfounded and baseless, as long as the business writer uses such jargon with professionalism in writing only to recipients who share the writer's knowledge of such "jargon."

These critics do not grasp—that sportswriters, columnists, reporters, novelists, and educators write with jargon just as frequently as business writers. When these critics advocate the use of popular common language as replacement for technical jargon, they seldom acknowledge that popular language is loaded with its own "jargon," which is incomprehensible to most people. (For example, even many who watch football games on television are now aware of the precise meaning of the term "fly-pattern.")

Professional usage of jargon sets apart the business writer as one who, through education, accumulated knowledge, and years of experience has acquired a learned vocabulary and terminology as the mark of his/her professionalism. It is a credit, a high mark; it gives the writer an underpinning of reliability.

## 4.2 Technical Terminology

Whenever writing, business writers should use as much of their discipline's technical terminology as is warranted to provide the recipient with a precise understanding of the subject matter.

Whereas the business writer needs to ensure sufficient comprehension of his/her text by the relevant recipient, this does not mean that the writer should shun using the precise technical terminology of his/her field. Indeed, substitution of simpler, more commonly used words for technical terminology (as advocated in some books) will more often than not result in ineffective communication, as well as in the relevant recipients forming an unfavorable image of or reaction to the writer.

The following brief examples will illustrate this point:

- "Direct costs" is a cost accounting term. Its precise meaning is known to accountants and non-accounting discipline educated businesspersons (for example, general managers of divisions) who nonetheless know its concepts and methods. "Direct costs" is an accounting term used only for direct materials and direct labor costs traced directly into specific units of products manufactured, assembled, or otherwise produced.

  When the financial analyst refers in the performance analysis report to the actual versus planned comparative results of "direct costs," he or she cannot substitute common English language for this term so that all recipients will understand the concept, even though some recipients might not be familiar with the precise meaning of "direct costs." The writer cannot substitute a more expansive language, such as "the costs which were directly incurred by the division for the purpose of manufacturing Product 'A,'" because this would have a completely different meaning to the recipients.

- Technical terminology often appears to uninitiated readers as mere code language, but the business writer should use such terminology whenever professionalism demands.

  "FIFO" and "LIFO" are conventional accounting terms that stand for the inventory cost accounting method ("costing-out") of first-in, first-out ("FIFO") and last-in, first-out ("LIFO").

  It would be unprofessional (and pedestrian) for the business writer not to use "FIFO" and to translate such a term in clearer English as, for example, "the cost of those products which were made earliest should be first written-off as costs from the Inventory Accounts..."

- A commonly and frequently used accounting technical term is "return-on-investment," most often written only as "ROI."

  There is no proper substitute for this term, since "return" connotes pretax or net profit and "investment" connotes what are specifics in each situation as the investment.

## 4.3 Special Terminology of Companies and Industries

The task of business writing is further complicated since business writers need to learn the terminology specific to their own companies or industries and use such terminology appropriately, regardless of how unfamiliar such terms sound or appear in writing to the uninitiated. The business writer who accomplishes this possesses a higher degree of professionalism than the one who does not.

A prime example is the term "bedrock expenses," which was created at ITT. This term found its way into the lexicon of electronics companies that looked at ITT's Financial Control and Planning system as worthy of emulation in the 1970s. This was a unique and precise term with regard to its concept. It basically means, in profit planning, the identification of expenses remaining to be incurred by the corporation even after shutting down and locking the gates of a factory or the company itself.

Also in the electronics industry, the term "yield" has acquired its own specific meaning, which cannot be substituted with any other word or common English description. Moreover, "yield rate" and "yield factor" became even more industry specific terms.

Even in the stock market and bond market, in which millions of American are involved, the simple words buy/sell and short/long are technical jargon.

## 4.4 Governmental Terminology (Jargon)

In the above sections we have discussed the technical terminology of disciplines (e.g., accounting), which are well respected by the public for their professionalism. It is safe to add that the public has also accepted such terms as "strong/weak" when referring to currency exchange and "long/short" when referring to buying and selling stock, despite these terms' obvious lack of clarity to the uninitiated person.

The public's criticism of governmental "bureaucratic language" rivals its criticism of lawyers for their excessive use of "legalese." Nonetheless, specific governmental terminology has evolved in each of the vast array of areas where one or another (or several) governmental departments, agencies, or sectors are involved. Setting aside such obvious terminology found with regard to laws and regulations, here the discussion is concerned with more subtle points about this terminology.

The business writer who writes on his or her company's behalf to a given governmental agency must possess knowledge of the pertinent governmental terminology when writing the specific letter or report. The difficulty of this task is multiplied for the business writer who holds a position in which he or she has to deal with business situations involving various governmental agencies, and is further multiplied if he or she has the responsibility of writing to such government agencies in foreign countries as well. Whereas knowledge of accounting, etc. terminology is important when writing to non-governmental recipients, it is essential that the writer know the government terminology relevant to the situation.

### 4.4.1 Terminology of Government Form (Manner)

Even the form (manner) of communication has specific "governmental terminology." For example:

- The Customs Service issues "rulings" to companies or to companies' agents or outside representatives for prospective or future import transactions. Thus, the business writer has to know and write in accordance with the Customs Service's terminology (jargon) specific to "rulings."
- However, when the business writer has to write in response to the Customs Service's "decision" which was issued about the company's import transactions, the Customs terminology (jargon) is different in certain respects than would be in the case of "rulings."
- In addition, the European Community's Customs Headquarters issues "directives," which has its own specific terminology. Malaysian, Japanese, and other Customs Services also utilize specific terminology, even for the names of their written communications (such as "notices," even though it is notice of decision and not public notice).

The importance for the business writer regarding the above is that he/she needs to know the meaning and relevance of such governmental jargon. If the writer doesn't specifically know the governmental terminology pertinent to each business situation about which he/she writes, then this individual should not be the one from the corporation to write to the government, regardless of his or her brilliant writing style, or persuasive, expository, and informative language.

### 4.4.2 Government Terminology

The following are but a few illustrative examples of governmental terminology which, but for the government, would not have been created or would have completely different meanings.

- "Dumping" is an example of special governmental terminology, and refers to the allegation and investigation by the U.S. Commerce Department of foreign products being sold in the U.S. at prices below those in their domestic foreign market, thus harming American producers of competitive products (i.e., "dumping these products on the American market").

  The term "dumping" is also found in foreign countries' governmental terminology. Yet, even to most lawyers the word "dumping" would be associative with chemical or other waste dumping.

  Thus, the business writer should be careful when using the word "dumping" if the subject matter of his letter or report involves international transactions. (Take, for example, the following statement: "Since there is no market to sell the rectifiers which failed the final Q.C. tests, your department should dump these off as scrap sales." While it was not the intent of the writer, such a comment can result in dumping allegations in the foreign country.)

- "Normal price" has a much different and defined meaning in governmental terminology than in accounting, marketing, or business in general. It is far more restricted in meaning than, say, "Normally, the price of our company's batteries is..." *Normal price* to the government official means the price at which such a product is usually offered in the industry to trade customers at "arms-length" at the same level of trade in the same quantities (and even "same" means only "about the same").

- "Administrative measures" is a term used by several foreign governments. While "administrative" has a reasonably reliable meaning to the business writer, he/she needs to find out what "measures" means in governmental terms in the specific country. This is particularly the case since in U.S. Government literature there is no specific connotation tied to this term, although it appears in general usage.

The reason for the focus here on governmental terminology is that the business writer needs to consider this area most carefully with regard to his/her writing, in order to obtain favorable results or to prevent exposures to his/her company from the governmental authorities. This adherence by business writers is easier said than done.

## 4.5 Pitfalls of Language in Global Operations

Assuming that the business writer has an excellent command of the English language, he/she must be extra careful in writing to foreign nationals within or outside the company.

This section only addresses aspects of language and terminology. Under Sections 5.4 and 5.5, the effects of variables of education, knowledge, experience, and cultural factors in business writing are further addressed.

First, when writing to foreign recipients, the business writer should disregard the conventional rule of business writing with regard to emphasis on conciseness or brevity. The writer should expound on words, terms, and meanings by interjecting explanations, such as: "This means...," "In other words...," or "To state this another way...." In other words, "spell it out."

Second, the business writer should assume the role of the foreign recipient by speculating upon which words, terms, and expressions might be unclear and result in their misunderstanding. It is always safer to underestimate the level and degree of their English language proficiency than to assume that they know the particular word or term used.

Third, despite the pitfalls of "language barriers," the business writer should not use simpler words, terms, or expressions than he or she would use when writing to American recipients. The solution is, as stated above, to provide a clear understanding by explanations.

Fourth, for the sake of clarity, it is most important in such situations to write using simple instead of compound sentences. Conjunctions (connective words) such as "but," "but for," "for," "yet," or "nor" which are clear to those competent in English may differ in meaning or be absent altogether in certain foreign languages.

Fifth, the business writer should abstain from using figurative expressions and metaphors which can be safely used to drive home or underpin points when writing to American recipients. For example: the statement "We should not throw ourselves at their mercy for an equitable determination of the (issue)…" enhances the writer's image in the minds of the president and legal vice president, but would probably confuse some of the foreign recipients. Such an expression will probably "go over their heads."

# CHAPTER 5

# The Business Writer and The Business World

## 5.1 Effects and Criteria of Business Writing

Often when the business writer addresses the issue of prevention or mitigation of losses or exposures to the company, he or she in fact has a "favorable impact" upon profits. In many other business situations, the business writer addresses matters in which only costs are to be incurred by the company, without benefits to "profit."

For example, one would have to stretch to draw parallels between writing about government regulations, employee benefits, or improvements in the factory's air-conditioning system in a tropical country and "profit." Another example is the business writer who recommends that the company obtain "typhoon insurance" coverage, which is a cost, unless a typhoon does cause damage to the facility.

It has been emphasized that the two criteria for correct business writing are: (1) correct English usage, and (2) clear, correct message content; the latter is deemed as *essential,* while English with regard to style, grammar, and mechanics needs to be at a sufficient level for the recipients to have a clear understanding of the written material. This is a unique statement, especially considering that this is a book on writing.

## 5.2 Communicating to Internal or External Audiences

The most basic classification of business writing is whether such writing is addressed to an "internal" or "external" audience.

When the recipient (audience) of the written communication is an employee of the same corporation or organization as the business writer, such communication is classified as written for any "internal audience." Here, the term "corporation" or "organization" is emphasized because even though the business writer from the New York corporate office may write to the corporation's Japanese group office or Malaysian subsidiary, it still falls under the category of "internal" communication, albeit the sender and the recipient are physically separated by over 8,000 miles. This is also termed "organizational communication."

When the recipients are not employees of the business writer's corporation or organization, then it is written for an "external audience," i.e., to "outsiders."

It is important to note in advance that the business writer has a different framework and obligation when writing "internally" or "externally." Aside from the differentials in subject matter, purpose, and content, the writer has to consider the style, formality, degree of decorum, the image of his/her company or organization projected through the writing, and the "tone" of the written communication, considerably more so when writing to an "external audience." These are essentials of "external" writing.

Actually, there is a wide range in each of the above factors for the business writer to choose from in writing to "external" recipients. For example:

- Writing to government officials at any level requires formality, decorum, a respectful tone, and extra attention with regard to the content, even when the writer has a close working relationship with the given government officials who are the recipients of the letter or report (seldom a memorandum).

This should be maintained even when the writer merely discusses in the letter exchange of opinions about general topics that are non-specific to the writer's company.

- On the other hand, when writing to outside public accountants, consultants, lawyers, or such service personnel as the manager of the freight forwarding company, the writer may elect to be more informal, forthcoming, and informative than in other "external" (or even than in some "internal") writings about similar subject matter.

All the factors which are discussed in this chapter are applicable in principle both to "external" or "internal" writing, although the level and degree will vary greatly.

## 5.3 Interdisciplinary Background Recipients: Finance, Marketing, Engineering, Law, and Others

Many times, business professionals write letters, memoranda, reports, or other forms of communication to recipients who come from "interdisciplinary backgrounds." The following example illustrates what is meant by this.

The Assistant Controller (who has an accounting degree) of the company writes a report to the organization's Financial Vice President, a subsidiary's General Manager (who has a degree in engineering), the Marketing Vice President (who has a degree in marketing), the Trade Affairs Director (an attorney), and the Distribution Director (a former Customs official).

These are referred to as different educational fields or disciplines (i.e., accounting, marketing, engineering, etc.).

Business writers are all influenced by, or governed largely by, their own fields or disciplines of education and expertise. Each discipline differs from other disciplines in terms of education, knowledge, familiarity, scope, concepts, terminology, techniques, methods, and jargon (a factor not to be lightly dismissed). In fact, even within the narrow field of "accounting" there may exist a wide disparity both in terms of breadth and depth of two business writers, albeit they are both "accountants." Exhibit 5.3.1 illustrates in a very limited manner the vast interdisciplinary diversity of writers and recipients.

This situation is not limited to Fortune 500 corporations. Such communication is often written in even small companies that have only a couple of hundred employees, since they also have diversified functional managers with different educational backgrounds. However, this factor is multiplied in the case of larger, more complex corporations, due to the further complications of different levels of recipients within the same functional departments.

For example, various reports may originate from different levels of writers within the Finance Department, issued by the Corporate Cost Analyst, the Controller, or the subsidiary's Planning Director to recipients at varying levels within the marketing organization (such as Product "A" Salesperson, Marketing Research Specialist, or Corporate Marketing Vice President).

### 5.3.1 The Writer Must Visualize the Recipients (Audience)

As an early task in the writing process, the business writer must consider the topic of the report, letter, memo, etc. and "visualize" the background of the recipients (audience) with regard to their education, knowledge, familiarity, or experience with each aspect of the writer's planned report.

Since this is the second time in this chapter that "education" and "knowledge" are both mentioned as factors, it should be noted here that the business writer needs to take into account both these factors regarding the recipient. If the writer does not know the educational and experience level of the recipient, he or she must make an educated guess as to these levels before proceeding further with the writing process.

The writer must bear in mind that in the business world there is a significant number of persons, including executives and managers, who have little or no college education. Conversely, other individuals may have advanced degrees in fields other than accounting or finance, but may have acquired a substantial knowledge or familiarity with accounting and finance through their work experience.

A good example of this case is the renowned Bill Gates, Chairman and CEO of Microsoft, who did not even complete his freshman year at Harvard. On the other hand, while most division and subsidiary general managers of Fortune 500 corporations are college graduates, only a small percentage of them have accounting or finance degrees, though all of them must have a sufficient comprehension of it—accounting and finance are the languages of planning and control in business.

For example, when the Corporate Financial Analyst writes a report that discusses the Malaysian Government's import licensing laws for "wire fabricating machinery," he/she should "visualize" what familiarity the Malaysian subsidiary's General Manager and the Corporate Associate Counsel have with this specific topic. The writer should not assume that these individuals are knowledgeable and experienced with import licensing laws in general and with those of Malaysia in particular. Neither the fact that they manage the subsidiary's operations in Malaysia nor that they are lawyers (even one experienced in international operations of the corporation) should lead the Analyst to such assumptions.

The above example may appear esoteric, but each year business writers issue millions of individual pieces of communication that warrant such "visualization" both for U.S. and for international business situations.

It should be understood that even when the business writer knows the recipient or has gathered as much information about the individual as possible in the Prewriting Stage, the business writer is merely left with his or her general idea of the recipient's degree of comprehension.

In summary, the business writer must consider ("visualize") the recipients with regard to the following factors:

- Field or discipline of education.
- Knowledge acquired in the field.

- Level and specialization of education or knowledge.
- Knowledge of the subject matter of the communication.
- Experience with the subject matter of the report.

## Exhibit 5.3.1
## INTERDISCIPLINARY DIVERSITY IN THE BUSINESS WORLD
## A PARTIAL ILLUSTRATION
## (REDUNDANCY)

ACCOUNTING
Cost Accounting and Control:
- Cost Accounting
- Cost Analysis
- Cost Estimating
- Cost Control

General:
- General Ledger Accounting
- Inventory Control
- Accounts Receivable
- Accounts Payable
- Cash and Funds Accounting
- Other

Consolidations Accounting
Financial Statements:
- Preparation and Reporting
- Analysis
- Government Reporting
- Other

Planning and Budgeting:
- Profit Planning
- Strategic Planning
- Operational Planning
- Budgeting
- Planning Analysis
- Budget Analysis and Control
- Other

MANUFACTURING
Production
Quality Control
Logistics
Industrial Engineering
Material Control
Material Handling
Manufacturing Administration
Standards and Measurements
Other

PROCUREMENT
Source Analysis
Purchase Orders and Controls
Price Analysis
Logistics
Expediting
Production Coordination
Material Handling
Other

Note: This exhibit does not even take into account the complexities of international operations.

### 5.3.2 Writing the Report for "Interdisciplinary" Recipients

When writing to multiple recipients, the writer has to conceptualize each part of the report and evaluate whether the *relevant* recipients will be able to understand it. The emphasis is on *relevant.* In cases of multiple recipients, the writer needs to prioritize or rank such individuals as to the importance of their understanding of the report's nuances.

Often the report must include precise terminology (jargon), formulas, methods, techniques, concepts, etc. The writer cannot expect all the recipients to fully (or even sufficiently) grasp every aspect of the report. The critical aim of the writer must be to ensure that the recipients who must understand and act upon the communication fully grasp all of the nuances relevant to them. The other recipients should be able to attain from the report a sufficient overall and basic understanding of the business situation.

The following example will illustrate the guidance in this book to business writers.

> Referring to the Malaysian import licensing laws for the "wire fabricating machinery" situation, let's assume that this machinery is essential equipment for the production of semiconductor products at the corporation's Malaysian operations. The Corporate Financial Analyst has the responsibility at corporate headquarters to investigate and explain within the corporation the movement of products and equipment to or from all foreign subsidiaries for their customs, import, export, and related government affairs—this is a highly specialized field of expertise.
>
> In addition to the Malaysian subsidiary's General Manager and the Corporate Associate Counsel, the report's recipients include the Corporate Manufacturing Vice President, the Corporate Export Manager, the Assistant Controller, and the engineer in charge of installing such equipment at the factories.

The writer in the above example must ensure that he or she concisely, but sufficiently, provides the information of the Malaysian import licensing laws as they pertain to the importation of the "wire fabricating machinery" with explanations and documentation for the precise understanding of the General Manager and Corporate Associate Counsel. Both individuals have sufficient knowledge and experience with import licensing as general principles; hence, they are expected by the writer (the Analyst) to understand the detailed nuances involved, as presented in the report.

Conversely, the Assistant Controller is responsible at corporate headquarters for cost accounting and analysis; the engineer's interest is in the installation of the equipment; and the Corporate Manufacturing Vice President wants the machinery for production purposes in Malaysia. The Corporate Export Manager needs to know only when, at what invoice prices, and with what type of documentation the machinery will be exported to Malaysia from the United States.

Thus, in this situation, the Corporate Financial Analyst properly omits explanations of the basics and nuances of import licensing laws, even though these other recipients may not be familiar with the subject itself. This subject is not relevant to them except for the consequences associated with it—they are *passive recipients* of the information. It suffices for them to understand from the report that in order to import this machinery, the General Manager and the Corporate Assistant Counsel have the responsibilities to comply with and obtain the permits to import this machinery under the Malaysian import licensing laws. Otherwise, this machinery will not be installed in Malaysia.

Conversely, the Corporate Financial Analyst's discussion of startup costs, return on investment, and transportation logistics are not relevant to the Corporate Associate Counsel and the engineer; hence, the writer does not need to explain the nuances of such concepts, methods, or terminology for their grasp. These factors are relevant to the Assistant Controller and General Manager (to a lesser degree of detail) and the Corporate Export Manager (i.e., transportation logistics).

Upon a cursory review, the above example may appear to be an exception, and thus an unusual occurrence in business. However, in Chapters 3 and 4 it was already presented that the principles addressed in this chapter apply widely to business situations, including those of small companies (albeit with less complexity). For example, in a small company with U.S. operations only, the Cost Accountant also would not need to explain in detail what was included in the costs of Product "A" in the monthly performance report to the President, Marketing Vice President, Controller, and Factory Manager.

Moreover, these principles apply equally to letters, memoranda, and reports written to external parties such as government officials, consultants, lawyers, public accountants, and persons in other companies. Nor are these writing principles restricted to "industry"—they are also applicable to department stores, banks, organizations, law and accounting firms, and government organizations.

## 5.4 Integrated Organizational Structure and Level

Writers of business letters, memoranda, reports, and other communication must first study and understand the organizational structure and levels of the company, governmental agency, and outside organizations to which they write.

Many companies have complex organizations that are divided into corporate, area operations groups, product groups, subsidiaries, divisions, branches, and sub-regional offices. Within these groups, the operation line and functional organization is extensive.

Exhibit 5.4.1 is a segmented illustration of a multinational's financial organization at the corporate and Latin America area group levels (the area group financial department being a mere extension of corporate finance, located in Japan in this exhibit). In actuality, this chart would be more extensive, both horizontally and vertically.

A useful safeguard for the business writer is to provide a copy to the organization's highest authority in the functional area of the recipient(s) to whom the correspondence is addressed or directed for response or action. For example, the Malaysian subsidiary's Procurement (Purchasing) Director should copy the memorandum in which he or she is requesting information of the sources and prices of wire for the rectifiers from one of the 18 Corporate Procurement staff members for the Corporate Procurement Director. The writer in this instance ensures that if the request was addressed to the wrong recipient, the Corporate Procurement Director is involved to channel it to the person responsible for wire procurement.

In such instances, the writer should include a brief cover note, indicating that the Director does not need to be involved with the subject matter but needs to ensure only that it is handled by the person responsible for it.

## 5.5 Decentralized and Global Operations

The above section discussed the business writer's task of considering the recipient's education and other factors for his/her comprehension of the report, letter, etc., as well as for the writer's determination of who the recipient should be in the particular organization (internal or external).

All of these factors present significant incremental problems that the business writer must carefully consider when writing in a decentralized or global business operations environment.

### 5.5.1 Writing in the Decentralized Operations Environment

The business writer who has carefully studied and understands the organizational structure and operations of his/her corporation also understands where the responsibilities and decision-making authority lie within the company. This understanding is an integral element of business writing.

This task is made more difficult for the writer when his or her corporation has implemented a decentralized management policy. That is, when most of the daily or detailed managerial decisions are the direct responsibility of the operational and functional managers of divisions, factories, or subsidiaries, instead of those at corporate or group levels.

In this business environment, the business writer has the additional burden of "visualizing" the comprehension capability of managers and staff at various distant or separate decentralized operations of whom the writer is most likely to be at best only vaguely familiar. This scenario is not at all uncommon. The Fortune 1,000 Industrials each have at least a dozen separate divisions, factories, and subsidiaries. Many retail businesses have dozens if not hundreds of branches, as do banks, insurance companies, and professional firms. And, the U.S. and foreign governmental organizations are highly compartmentalized.

Using the example of the Malaysian import licensing situation from Section 5.3 for unity and consistency, we will retain the writer as the Corporate Financial Analyst but substitute the written communication with an investigative memorandum. Here, the Analyst writes to the Malaysian subsidiary's Controller, Import Manager, Manufacturing Director, and General Manager to obtain their report as to what factors would be involved if the "wire fabricating machinery" is exported from the United States to their factory.

The terminology, concepts, principles, potential problems, laws, and regulations need to be explained, posited (posed), and discussed by the Corporate Financial Analyst in a much different manner than under the circumstances of his or her informative report to the previous different set of recipients. In some aspects, the subsidiary's personnel are more knowledgeable or experienced, while in other aspects they lack the conceptual or fundamental knowledge.

This example also illustrates the vast difference in the purpose of the business writing as well. Here, the writer's purpose was to provide information. To a large degree, the purpose governs what the writer includes and how it is included in the letter, memorandum, report, or presentation.

The business writer's task becomes vastly more complicated when writing to decentralized management since he or she has different situations to address to personnel at several widely differing decentralized divisions, factories, and subsidiaries. As a further complicating factor, he/she often has to include copies for persons at the corporate or group levels, which requires his or her "visualization" of what these remotely involved persons need to know, from the correspondence.

The business writer employed in these "service" organizations (as compared to "industrials") needs to apply the same diligent Prewriting Stage review, research, investigation, evaluation, and organization for writing for the given business situation as in the "industrial" business world. He or she needs to carefully select the recipients of the correspondence and prepare it for their needs as well as his or her own.

### 5.5.2 Writing in the Global Operations Environment

The business writer has to bear in mind for all of his/her writing that governmental issues or factors may be involved, be they U.S., local country, or other (e.g., European Community or World Trade Organization). He/she has to write accordingly, covering a greater scope of issues than when only writing intra-U.S.

## 5.6 Multilingual and Multicultural Considerations of Audience (Recipients)

When the business writer's recipients (audience) include persons from foreign countries whose native language is not English, he/she has the additional task of ensuring their comprehension of the correspondence.

First, this requires that the writer either write at a simplified level of English or that he/she includes additional explanation of words and expressions that the foreign reader might not otherwise understand. The writer must again "visualize" his/her audience, putting himself/herself in their place, and guess what terms in the correspondence may need clarification. The experienced international business writer always emphasizes to the recipients the necessity of their requesting clarification of anything they do not clearly understand.

Second, the writer must consider not only that his/her correspondence is clear to the recipients in terms of English, but also that such technical accounting, engineering, marketing, etc. terminology has the same meaning to the recipients as it does to the writer.

Under this criteria, it is not meant that the writer considers the technical education, knowledge, or experience of the recipients for such elements as the accounting concept of "direct cost" method or "return on investment" formula. Rather, it is meant, for example, for such terms as "compute(d)," "tier," and "interrogatory." The business writer needs to ensure that the meanings are the same to the recipients as they are to the writer, even if further explanation adds to the length of the text. An effective method for explaining potentially unclear terminology is to attach an addendum (like a glossary) of such terms at the end of the correspondence.

Many business writers are under the misguided assumption that the solution for overcoming such potential misunderstanding on the part of foreign (and even fluent English speaking) recipients is to simplify the language in the correspondence, and omit technical terminology or substitute as much of it as possible with simpler wording.

This should be avoided by business writers at all times. Aside from the need for precise immediate understanding on the part of the recipients, the writer is educating and building a better base of communication for future correspondence with the same recipients. ("Interrogatory" can be explained but not substituted with another word; "commuter value" in customs terminology cannot be substituted by "calculated value.")

Further, it is in the arena of international written communication where the business writer can and should show his/her learned knowledge of appropriate foreign vocabulary and terminology. (In a more restricted scope, he/she should do the same in intra-U.S. written communication, but his/her image will be more elevated in the eyes of foreign readers by such learned language.)

Here, we have included Exhibit 5.6.1 to list a few foreign words and phrases that provide the image of cosmopolitanism on the part of the business writer who properly incorporates them into his/her writing.

The business writer needs to grasp and pay attention to the manifold and often subtle multicultural elements in global written business communication. The following are only a few of the multicultural aspects that the business writer should keep in mind:

- Employees of the business writer's company in many foreign countries feel that they are much more under the control of their governments than do Americans. Therefore, the business writer has to be more careful and guarded in what and how he/she writes, in order to ensure that the foreign recipients do not misconstrue something in the written communication as being an infraction of a particular government policy, law, or regulation. An employee is likely to discuss it informally with a government official, which often results in formal scrutiny.

- In many foreign countries, the social structure is highly hierarchical, and in foreign companies and in American companies' foreign subsidiaries the same culture prevails. Therefore, the business writer should be very careful to "write to the top" of the departments (marketing, engineering, finance)—i.e., the Director or Vice President of the department, or at least always copy them when writing directly to a staff member.
- The business writer should never write directly to a department staff member, but write only to the department head (like the Controller) when there is actual or implied criticism of the department included in the written communication. The "loss of face" syndrome in many foreign countries, as well as actual harm to the prestige of the superior in the department, should cause the business writer to follow this as an unwritten rule.
- The business writer needs to carefully edit the written communication for language that is accusatory or insinuative in nature, or such language that is prone to misinterpretation and might result in emotional reaction by the recipient, which was *not* the writer's intent.

## Exhibit 5.4.1

## MULTINATIONAL CORPORATION FINANCIAL ORGANIZATION

## CORPORATE AND LATIN AMERICA GROUP

## A PARTIAL LIST BY FUNCTIONAL RESPONSIBILITIES

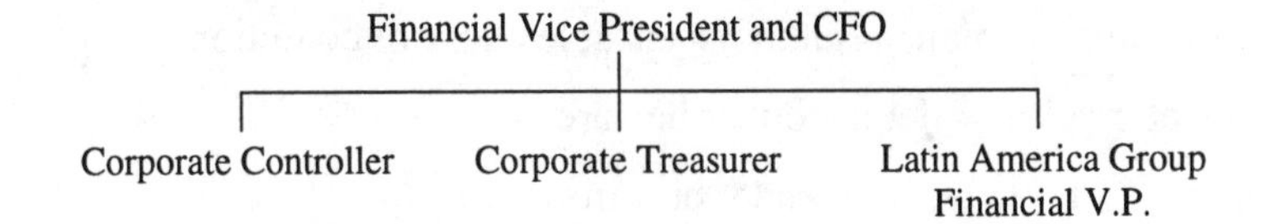


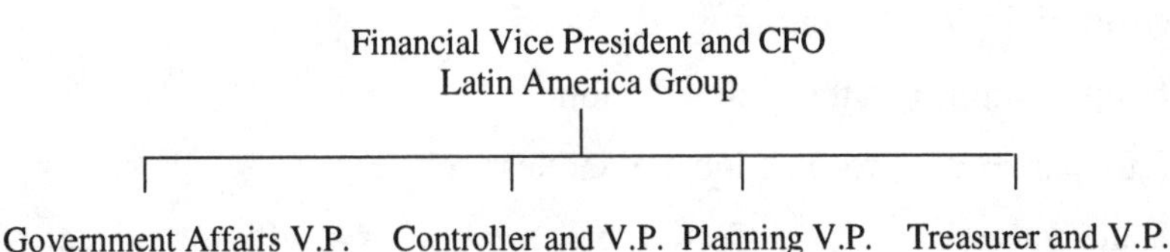


| Government Affairs V.P. | Controller and V.P. | Planning V.P. | Treasurer and V.P. |
|---|---|---|---|
| • Export/Import Controls | • Fin. Statements | • Profit Planning | • Investments |
| • Customs | • Fin. Reporting | • Strategic Planning | • Currency Controls |
| • Gov't Reporting | • Assets Control | • Financial Analysis | • Banking |
| • Foreign Investments | • Procedures and Manuals | • Budgeting | • Cash and Funds |
| • Gov't Relations | • Cost Control and Analysis | • Feasibility Plans | • Stock Market and Stockholders |
| • Other | • Audits | • Other | • Other |
| | • Accounting Controls | | |
| | • Pricing Controls | | |
| | • Other | | |

## Exhibit 5.6.1

## FOREIGN WORDS AND PHRASES APPLICABLE TO THE BUSINESS WRITER

ad hoc = for a special purpose

ad referendum = for further consideration by one having the authority to make a final decision

causa sine qua non = an indispensable cause or condition

caveat emptor = let the buyer beware

caveat lector = let the reader beware

de minimis non curat lex = the law takes no account of trifles

enfin = in conclusion

fait accompli = a thing already done

faux pas = an error in etiquette; tactless act

haute vulgarisation = effective presentation of a difficult subject to a general audience

homme d'affaires = man of business; a business agent

hors commerce = outside the trade; not offered through regular commercial channels

modus operandi = method of operation

per se = by itself

post hoc, ergo propter hoc = therefore on account of it (a fallacy of argument)

quid pro quo = something for something; an equivalent

secundum artem = according to the accepted practice of a profession or trade

sine qua non = essential condition; without which nothing

status quo = condition in which it is

verbatim = word for word

vis-à-vis = face to face with; opposite

# CHAPTER 6

# Types of Business Writing: Internal and External

## 6.1 Overview of the Types of Business Writing

In this chapter, the main types of business writing, ranging from letters to abstracts, are discussed at some length. Generally, all writing in business, in government, and in professions, such as law and public accounting, are included under the umbrella term of business writing. However, as noted in Chapter 1, business writing includes a narrower scope of writing based on the main criterion of "purpose" of the written communication.

An enumeration of the types of business writing covered in this chapter is helpful at this point:

- Letters
- Memoranda
- Reports
- Proposals
- Plans
- Presentations
- Procedures, Handbooks, and Manuals

- Abstracts and Reviews
- Responses to Communications Received

The main, governing criterion for grouping Business Writing is whether such writing is "internal" or "external."

"Internal" writing takes place when the writer writes to a recipient who is employed at the same organization, corporation, government, public accounting firm, think-tank, consulting group, or association. Stated another way, this writing is "intra-organizational."

"External" writing takes place when the writer writes to a recipient who is not employed *in* the writer's organization (emphasis is on *in* as opposed to *by,* as clarified below).

This grouping of business writing by "internal" and "external" governs these writings in various ways, as discussed in this chapter and in other parts of this book. Simply put, the business writer from Corporation ABC has a list of "rules" and convenants as to why he or she has to write differently to employees of Corporation ABC than to recipients who are not employees of the corporation.

The following pertains to all types of business writing.

- While the business writer from the corporate office and the recipient at the Malaysian subsidiary and Japan-based Asian group office are employees of the same "corporate family," they are not as closely *related* as the business writer is to the corporate office's employees.

  This point is important to the business writer for his or her evaluation as to what should be included on certain sensitive matters, and how such matters should be presented to recipients of Corporation ABC who are, however, not corporate office employees. (For example, if at the corporate office the CEO and a small group of executives and expert staff are evaluating the relocation of production from Malaysia to Mexico, the business writer has to differentiate very carefully what to include in his/her writings to the various recipients. These are *purpose* and *audience* factors.)

- Regarding the above point and example of the business situation, the business writer at times includes in the letter or report data, information, or discussion to "external" recipients (lawyers, CPAs, government officials, etc.) which he or she reveals only to a few "internal" employees.

  Thus, the "internal" and "external" grouping at times is blurred. Moreover, this can represent a dilemma to the business writer from managerial and even legal standpoints.

- Books on business communication/business writing usually state that writing is more formal, restricted, or reserved for "external" recipients, including in form of address and tone. This is valid for the examples usually provided for employment, collection, credit, damage claims, inquiries, and résumés.

- However, as discussed in this chapter and elsewhere in this book, in writings which are classified as "business writing," the business writer often writes less formally (including addressing by name), and in a less restricted and less reserved manner to an "external" recipient than to certain "internal" recipients. This may be the case, for example, in writing to outside lawyers, consultants, or public accountants who are employed (retained) by the company where the writer works.

The above limited background discussion should be helpful to the following detailed discussion of letters and other types of business writing. Moreover, the material presented and discussed in the previous chapters of this book are integral parts of, and are pertinent to, the following types of business writing.

## 6.2 Letters

Generally, business letters are a form of written business communication to "external" recipients (audience). The conventional and acceptable form of proper business letters, as described below, renders the letter format applicable to "external" written communication, and usually inapplicable in writing to members of the organization of the writer.

Nonetheless, in numerous business circumstances and situations, primarily but not exclusively when writing to foreign group or subsidiary senior officials or technical personnel of the writer's own organization, the more formal business letter is used instead of the less formal "internal" memorandum.

It is customary in business letters for the writer to state why he or she is writing to the addressee in the *opening* sentence or brief paragraph. Since the writer represents the company, the use of "we" (i.e., the company) may be often more proper than "I" (to refer to the writer); however, this pronoun should be used appropriately. Examples of openings are:

- "I am responding to your letter of February 10, 1996..."
- "We are submitting this application letter and attachments for your kind consideration for approval of..."
- Ms. Jane Smith asked me to respond to your January 8, 1996 proposal concerning the feasibility study..."

In the case of a positive, favorable, or beneficial (at least in the writer's mind or opinion) recommendation, conclusion, or decision to the recipient, the writer should use *deductive order* of arrangement in his/her letter. Here, the writer presents in the introductory part of the letter such recommendation, conclusion, or decision which also grabs the attention of the recipient, due to his/her benefit being projected up front. Then the writer presents in the body of the letter the explanation, information, or arguments which support such a recommendation, conclusion, or decision.

Conversely, the writer should use the *inductive order* of arrangement and place in the conclusion of the letter any: (1) recommendation for action on the part of the recipient, (2) request for a decision by the recipient which is primarily beneficial to the writer's company, or (3) advice to the recipient of a decision made by the writer's company which is negative to the recipient. Then the explanation in the body of the letter serves as a buffer or provides the understanding to the recipient for rendering a decision or action favorable to the writer's company.

In the closing part of the letter, the writer should suggest or request a response, decision, or action by the recipient, or advise the recipient of the writer's or his/her company's decision or action.

It is businesslike, and usually builds goodwill, to end the letter on such as note as: "We look forward to hearing from you at your earliest convenience," "Thank you for your consideration of this matter, which is of great importance to us," or "We hope that this information will be helpful to you."

Ending the letter with "We regret...," "We are sorry...," or "We wish your success..." in the case of a negative decision by the writer or the writer's company towards the recipient usually sounds trite, insincere, or even like a standardized perfunctory cliché. Frequently, such phrases create a feeling of dehumanized, impersonal, and glossed-over treatment by the writer, or by the person(s) on whose behalf he/she writers the letter, of the subject matter which is of importance to the recipient. Business writers should avoid this pitfall. The same message can be sent in a subtle and courteous manner within the letter itself.

In this section, inductive and deductive sequence are discussed as they pertain to letters only, since letters are uninterrupted units, read by the recipient without the aid of separation by topical headings, as is used in memoranda, reports, or proposals. An effective substitute for such headings is underscoring, boldfacing, or italicizing key words or sentences in the text.

### 6.2.1 Format and Style of Business Letters

All business letters should have the following format elements:

Letterhead: Each company, organization, association, and firm has its own letterhead printed on its stationery, forms, and other vehicles of correspondence (e.g., fax cover sheets).

Dateline: The date of the letter should always appear written out under the letterhead, such as:

January 10, 1996

It serves no purpose to abbreviate the date, such as "Jan. 10, 1996" or "1/10/96." (The latter can be mistaken in foreign countries where day, month, and year is the proper order—1/10/96 meaning October 1, instead of January 10.)

Inside Address:

Name
Title
Company Name
Street Address
City, State, and Zip Code

In letters to foreign countries, the "Street Address" line is followed by the "City, Country" line. (For letters to governments, see note below discussion of "CC:" section.)

Salutation: In letters addressed to Americans in the United States, the following are appropriate forms of salutation:

> Dear Ms. Brown:, Dear Mr. Commissioner:, or Dear Senator Jones:

If the writer is on familiar terms with the addressee, then "Dear Bill," for example, would be an appropriate salutation.

In Chapters 5 and 7, the decorum, protocol, and etiquette in correspondence are covered in detail, including relationships to the international environment.

Message: This is the body of the letter, which is further discussed in detail in this chapter.

Complimentary Closing: "Sincerely," or "Sincerely yours," is the proper way to close most letters.

In letters to government officials or senior executives, the writer may choose the more formal closing of "Very truly yours," or "Respectfully yours,".

When the writer has a close personal relationship with the addressee, or wants to project a familiar tone, then "Regards," or "Best regards," is acceptable.

If the letter is quasi-personal in content, such as a congratulatory letter for the recipient's promotion to a new position, or one of well-wishing for a new business venture, then the business writer may go as far as using "With warmest regards," "With best wishes," or "Always with best regards,".

Glitzy closings should always be avoided—if the business writer wants to express such a message, then it should be in the message (text) of the letter.

Signature: In "external" letters, the signature bloc should be included, such as:

James Smith
Vice President
Manufacturing Operations

The letterhead reveals the company or division, subsidiary, or other unit of the company of the writer. The writer signs his name, preferably legibly, above the signature bloc.

If a staff member or the secretary signs the letter in place of the writer, then it should be done as follows:

(Person's Signature) for
James Smith

Enc. or
Enclosure: If there are enclosures or attachments to the letter, then the word "Enclosures" or the acceptable abbreviated form "Encs." should be typed below the signature bloc's last line. The writer may choose to enumerate the enclosures, such as:

Encs.—(1) Copy of Malaysian
Licensing Law

(2) Copy of Smith, Jones
consulting advisory

CC: When the writer includes other recipients beside the addressee for the distribution of the letter, then their names must be listed along the line marked "CC:", which stands for "copy or copies to."

Dependent upon the nature of the letter, it may be a list of names in alphabetical order.

CC: John Brown, Nicholas Heath,
Ann Whitman (this can be
arranged vertically as well)

In some cases, each recipient's title and the name of their organization should be listed in the "CC:" section of the letter. While this is more applicable to "external" correspondence, it may also apply to "internal" letters, where the recipients are at different business units of the company. In these instances, the appropriate method is to list them in order of importance of title, not alphabetically by name.

A note is useful here about the "Inside Address" when writing letters to the U.S. Government. Although it is acceptable to use the method shown above, it is more proper protocol to use the following method:

Regional Commissioner of Customs (No name, even if
United States Customs Service known to writer)
(Name of) Region (e.g., "New York")
(Street Address)
(City, State, Zip Code)
Attn: (Title) (e.g. "Regional
or Counsel")
Attn: Thomas Taylor, Esq.
Regional Counsel

In such an instance, the salutation is optional to the writer. For instance he/she may choose: "Dear Mr. Commissioner," "Dear

Regional Counsel," or "Dear Mr. Taylor."

In fact, when the business writer encounters a situation when writing to a company, organization, or professional firm (e.g., law, CPA, or consulting) that he or she cannot find out the name of the cognizant person or even the title of the applicable staff member, it is acceptable to prepare the inside address in the above "governmental" format:

Company Name
Street Address
City, State, Zip Code
Attn: Controller ("Marketing Vice President," other)

In such situations, the proper salutation is either "Gentlemen" or "Dear Sir"; "Dear Sir or Madam" is a more recent non-sexist salutation recommended in some books. Never use "To Whom It May Be Concerned," because such a salutation gives the impression that the business writer does not have a clear purpose and is fishing to have someone interested in the letter. (Usually such letters are discarded, or at best forwarded by the mailroom to persons whom the mail clerk interprets as applicable.)

The three main formats of letters in American business usage are as follows:

- The traditional semi-bloc format, is moderately indented in the opening lines of each paragraph.
- The full-bloc letter is the "modernized" style, in which all lines begin at the left margin, with no indentation at all.
- The modified-bloc letter is the same as the full-bloc letter in all respects, except that the dateline and the complimentary closing line begin at the center of the page.

The traditional indented format is more proper in letters written to foreign addressees, because it is the format of correspondence in foreign countries.

Since letters are almost always written as "external" communication, the range of nature, purpose, and subject matter is immense for business letters.

When letters are addressed to "external" recipients, the business writer should be extra careful to adhere to all aspects of business writing presented in this book. The business letter is an image of—and reflection upon—the writer's company or organization, while the internal memorandum or report usually remains "internal" only, and reflects only on the writer or his/her department or division. For this reason, particular attention and detail were provided in the above discussion of letter format and style.

In the preceding parts of this chapter, business letters are treated as the medium for *external* business communication by the writer. However, the writer may find it applicable (or required by corporate policy) to write letters instead of memoranda to persons affiliated with the company or even employed by the company.

In addition to the cultural aspects of decorum discussed, the letter format is applicable when the writer wants to project an "arms-length" relationship between the company and the recipient. For example, the Human Resources Vice President writes a letter (instead of a memo) to the prestigious but titular Thai Chairman of the company's Thai joint-venture company about the company's consideration to implement a performance bonus system to increase productivity. The writer assumes that the Thai Chairman will feel obligated to show or submit this correspondence to certain Thai Government officials, and therefore the writer selects the *external* appearance of the letter, which also enables the writer to write with a decorum that underscores the high esteem of the company for the Thai Chairman for his prestige in dealing with the Thai Government officials.

### 6.2.2 Content of Business Letters

In addition to the above-referenced principles, the business writer should bear in mind foremost the purpose of the letter, the probability of the recipient understanding the subject matter expressed therein, and the recipient's potential reaction.

One of the most important incremental considerations for the business writer in "external" letters is to visualize and conceptualize the "external" recipient's breadth and depth of knowledge, familiarity, and comprehension of the letter's subject matter.

The business writer should rather err by providing more detailed and extensive explanations of issues, concepts, data, and other factors than by assuming knowledge or familiarity on the part of the "external" recipient. A prudent technique is the use of helpful attachments (addenda), instead of creating a letter that is longer than the normally conventional two or three pages (although some letters may require even five pages).

Tone is considered an essential of business writing. Note here that "tone" is much broader in scope and impact than the mere reflection of attitude, mood, and aim for positive reaction by the recipient. "Tone" in this more restrictive sense takes on added importance in "external" letters, because the writer speaks on behalf of his or her company or organization to outsiders.

The subject matter and the purpose govern the structure of the business letter. In general, all letters can be viewed as having the following main parts:

- Opening
- Introduction
- Body
- Conclusion
- Ending

In fact, in larger companies and organizations the business writer is largely guided by an operations handbook, a company procedures manual, or a similarly titled handbook or manual with regard to aspects such as correspondence format and style. Alternatively, the secretaries will know the company's policy or common practice about such secretarial matters, and the writer will acquire this from precedent and by experience.

The business writer is cautioned, however, not to indiscriminately rely on prior written material as "precedents." While this pertains more to content, it is also relevant to elements of format, where deviations are warranted, ranging from the inside address to the complimentary closing selection.

### 6.2.3 Multipurpose Letters

Business letters are addressed to a single individual. The title, position, and authority level of the recipient ranges from comparatively low levels to the highest levels in their companies, organizations, or associations.

The experienced business writer, however, considers when writing the letter which other functional persons will be involved with the letter at the recipient's company. Therefore, when preparing the letter, the writer should target his/her message to these other anticipated recipients as well.

For example, a letter is addressed by the writer, who is the Quality Assurance Director of his/her company, to the Sales Vice President of Smith Manufacturing Company ("Smith"), a supplier of rectifiers. In the letter, the writer advises the Sales Vice President about the failures of quality control (QC) tests of the latest lot of rectifiers received at the company's North Carolina factory.

The Director writes the letter and includes documentary evidence with QC technical language, terminology, and explanations aimed at the understanding of his/her Quality Assurance Director counterpart at Smith, since the QC failures are the reason—the heart of the matter—for writing the letter.

At the same time, the writer also targets the letter to the Sales Vice President for the purpose of obtaining replacements for the failed rectifiers. By extension, the writer also states or suggests that the Accounting Department of Smith must issue a credit memo or a "no charge" invoice for the replacement rectifiers, optional to them (this demonstrates "good tone" and a courteous approach in the letter).

To illustrate the complexities and multipurposes of business letters, the Director—at the instruction of the Controller, which the Accounting Department quantifies—may also submit in the letter a request (claim) for reimbursement of certain costs incurred by the company due to the substandard rectifiers (such as production costs in the products prior to realizing the problems with the rectifiers). This part of the letter and the supporting documentation is intended for the consideration of Smith's Accounting Department to issue the appropriate credit memo, or to question or even dispute the claim after Smith's internal review of the data.

In the above example, the letter is:

- A professionally written advisory letter to Smith about its QC failed rectifiers.
- A request (claim) letter for monetary reimbursements or replacement products.
- A request (claim) letter for reimbursements of costs incurred because of the substandard Smith rectifiers.
- A goodwill letter because the writer provides a detailed technical explanation in neutral, professional language, which is helpful to Smith's QC and other staff, instead of a burden upon them to have to analyze the problem without knowing what the writer provides to them in detail.
- A bad news letter for Smith about the problem with the failed rectifiers.

Here, the writer also wrote in good tone by providing the technical and financial information to Smith's cognizant personnel, instead of merely notifying them of the failed rectifier (bad news) and claiming reimbursement without providing them with detailed explanatory documentation of the incremental costs of his/her company, as caused by these rectifiers. Moreover, this letter pre-empts Smith's QC and Accounting staff from analyzing the issues without having the facts, information, and explanations presented to them, and arriving at conclusions resulting in conflicts between the two companies. (Disputes about the facts, information, and explanations result in businesslike disagreements, not in conflicts.)

### 6.2.4 The Business Letter Report

In rare business situations, the business writer may wish to use the business letter as if it is a report to the recipient.

Usually, the main reasons for electing the letter report format instead of the letter format are:

- The writer is able to present the subject matter in a clearer and more understandable manner than in a letter, due to the separation of parts of the letter report by appropriate headings, as is customary in reports.
- A letter is required but it would be inappropriate to attach a separate report to the letter, or the writer believes that the report as an attachment to the letter would be overlooked by the recipient as secondary material.
- Perhaps the foremost reason is, however, that reports project the image of objectivity of facts and information. If exactly the same material is included in a business letter, written as a letter, then the recipient will look for the writer's persuasive or convincing motives and language, searching for the writer's lack of objectivity or angle.

The differences between "letter report" and the regular report are mainly in appearance. An illustration of a letter report follows:

June 20, 1996

Ms. Sylvia Jones
Marketing Vice President
Smith Manufacturing Company
78 High Street
Stamford, CT ________
RE: Quality Control Performance of Smith Rectifiers

Dear Ms. Jones:
(Text with topic headings)

I hope that the above information will be helpful to your Manufacturing and Quality Control Departments in identifying the causes and correcting the excessive rate of failed rectifiers.

If we can assist you, please do not hesitate to let us know.

Sincerely yours,
Ed Parker

Quality Assurance Director
Encs.
CC:...

It is noteworthy how the business writer of the above "letter report" was able to include his "goodwill" and courteous complimentary closing, as if in a letter. This is not acceptable in a normal business report.

## 6.3 Business Memoranda ("Memos")

The business memorandum is the main medium for *internal* written communication *within* the organization (as opposed to the letter for *external* written communication).

It is noteworthy that the word "memorandum"—or as it is generally referred to in its abbreviated form, the "memo"—evolved from its associative root idea and from the reason for it: to reduce to paper information, ideas, principles, etc., instead of relying on memory for oral communication. In fact, the legal definition of memorandum is: "to be remembered; be it remembered... embodying something that the parties desire to fix in memory by the aid of written evidence..." In layman's words, the purpose of the memo is to "put it in writing" or "put it on paper."

In today's complex, regulatory, legalistic, interpersonal, and geographically distanced business world, almost everything is (or should be) "written down" for one's own "note for memo" (or file) or as "memoranda" issued to others within the organization.

"Reduce it into writing," "write it down," "go on record," and "put it into writing" are in fact rules, not just clichés, in the business world.

In this book, business writing is discussed from the aspects of persons who write communication for the purpose of affecting their organization's business affairs. Therefore, the regulatory, legalistic, and self-protectionist requirements for the "memo" are not discussed here, except in the context of the business memoranda. Suffice it to say about these aspects that the business writer should generally consider in his/her memoranda (as well as in letters, reports, and even notes) the regulatory and quasi-legalistic aspects involved, and avoid transgressing these.

### 6.3.1 Format of the Memorandum

Memoranda, just like letters, have a well-established, commonly used format in the business world. Regardless of their length, memoranda usually appear as:

DATELINE: The date of the memo should always be the first line below the letterhead. While it is best to show the full date, such as April 5, 1996, the date is often shown as 4/05/96. (As stated in Section 6.2, this can result in misunderstanding when the recipients include persons in foreign countries.)

FROM: The writer's full name (or in some company cultures first and middle initials and full last name) is shown, with or without his/her title. For example, John Talbot, Corporate Financial Analyst.

TO: The name of the addressee is shown, often with his/her title, and dependent upon company etiquette or practice, Mr. or Ms. in front of the name. For example, Ms. Mary Smith, Marketing Director, Pharmaceutical Products Group. If there are two or more addressees of the memo, their names should be listed alphabetically or by rank in the organization. (While Messrs. is proper to use in front of the names, it should be omitted if one or more of them are women.)

SUBJECT: A brief and precise title of the memo's subject should be shown. For example, IMPORTATION OF WIRE FABRICATING MACHINERY INTO MALAYSIA is an example from the Chapter 5 business situation.

ATTACHMENTS: The attachments should be briefly listed at the end of the memo, below the signature.

In the memorandum, there should be no complimentary closing nor signature bloc line, but the writer should sign his/her name after the last line of the memo to indicate that he or she read it and "signed-off" on its content. Further, the writer should not use letter-type endings, such as "Looking forward to hear from you," or "I hope that the above is helpful." The recipients of the memo work for the same organization; they are expected to respond to, act upon, and certainly read the memo as pertinent information to them (partially or in its entirety).

The business situation or subject matter governs the length of the memo. It can range from a couple of paragraphs to several pages.

The memorandum in Chapter 5's Malaysian import of machinery example is multipurposed. It requires a several page memo from the Corporate Financial Analyst, in addition to documentary attachments and schedules. It is also required that the writer structure the memo by its parts under appropriate headings, such as:

I. THE MALAYSIAN IMPORT LICENSING LAWS AND REGULATIONS

II. ALTERNATIVE PRODUCTION PLANS

III. VARIABLE PLANNED PRETAX PROFIT RESULTS

IV. RECOMMENDED ACTION

There are several other possible arrangements of headings by the writer in this memo, as there are in most memos.

### 6.3.2 Writing the Memorandum

The fundamental principle for the business writer is the same for memos, reports, and presentations as was previously discussed—either a deductive or inductive approach is utilized when writing the communication. Further, the writer must provide in the opening part of the memo a concise but clear understanding of the subject matter, problem, issue, and purpose for the memo.

It is also helpful to the recipient that the writer states in the opening part if the purpose of the memo is to provide information only; if it includes the writer's interpretations, explanations, or recommendations; if it requires specific action or decision by the recipients, or other points. Then the recipient can read the memo with an understanding of the writer's mind-set and purpose—including a more cautious and critical reading if the writer provides recommendations.

It is important to repeat an earlier piece of advice from this book that is especially relevant to the internal memorandum: the business writer should write the memo only when he/she has the information sufficiently complete to write it. Often, however, business situations warrant that memos be issued even when the writer has only partial information available, but such partial information should be sufficiently complete, and the writer should advise the recipient of this factor in the memo, explaining what information is missing from the memo.

The well-prepared memorandum is a total, self-contained unit, with its attachments and/or references to other written communication. The business writer should not prepare an abstract or summary as the executive summary of the memorandum (whereas it is often necessary or prudent to prepare such in reports).

## 6.4 Business Reports

It is impossible to define the term *report* even to the degree of the term "memorandum," which, itself, can be defined only vaguely. The reason behind this is that whereas the memorandum is a form of written communication "from" the writer "to" the recipient, which lists the "dateline" and "signature" in its format, the term "report" includes a vast variety of written "messages," often without even listing "from," "to," "dateline," or "signature."

Internal memoranda can be safely viewed as being informal reports. However, a main differentiation is that the reports should always be objective communications of factual information, even when the report does include interpretation or analysis. Reports lack the writer's persuasive, inducible, recommendatory, altering, or other influencing messages that are features of many business memoranda.

Stated in other words, the report is intended to provide factual information—to report information—with neutral objectivity. This point is of significant importance since the business writer must issue various types of reports; regardless of position, everyone in business needs to issue "reports" in some form. But, it is a common trait of writers to "try to be helpful" and add his/her input, even if just a little, in addition to the facts, so that the recipients can "understand it better."

In fact, the overwhelming majority of what are called "reports" do not involve the complete "writing process" at all—and the business professional should refrain from converting such reports into writing for his/her purpose.

**Exhibit 6.4**

**TYPES OF REPORTS COMMON AT CORPORATIONS**

**A PARTIAL LIST**

**General Manager's Office**
Monthly Manager's Letter
New Product Plan
Performance by Receptions
Government Issues
Monthly Intelligence Data
Manufacturing Department
Production Volume
Quality Performance
Productivity Rate
Startup Product Performance
Capital Equipment Requirements

**Procurement Department**
Purchasing Source Analysis
Purchases List

**General Accounting**
Inventory Status
Capital Equipment (Fixed Assets)
Cash-Flow
Payables Aging
Import Duties
Direct Labor Costs
Direct Material Costs
Variances From Plan

**Cost Accounting**
Manufacturing Costs by Product
Overhead Expenses
Administrative Expenses
Marketing (Selling) Expenses

### 6.4.1 Classification by Characteristics of Reports

The following basic classification of reports by their characteristics is helpful to the business writer:

Formal/Informal Reports. Formal reports are structured along the lines of established standards and conventions, and cover weighty business issues. Formal reports by their nature tend to be lengthy reports, while informal reports have no "prescribed" format or structure, and are comparatively simpler and shorter.

Information/Analytic Reports. Pure informational reports merely report facts, data, and events as they occur. Pure analytical reports provide objective interpretations and analyses of facts, data, events, concepts, plans, policies, or other issues/problems—the domain of the business (or technical) writer.

Internal/External Reports. This basic classification pertains only to the level of the report issuer's intent as to providing the report to members of his/her organization (internal), to outsiders (external), or to a mixed audience (internal and external). If the report issuer's intent is to write an internal report, but a recipient turns it over to an external party, such a report is still classified as internal due to its intent.

### 6.4.2 Informal Reports

All types of reports that lack the "formal" report structure presented in Section 6.4.4 below are informal reports, regardless of their weight of importance, language, or purpose.

We include under the informal category as well the important "periodic" reports: monthly, weekly, or other regular internal scheduled reports on manpower, division/product/company financial statements, capital expenditures, cash flow, etc. Such reports are informational and informal, even if the business writer submits these attached to his/her overlaying cover memo or letter (to outsiders). In fact, all informational reports are informal reports.

In informational reports, the criteria is to report the *factual* information as facts, data events, results, etc. that occurred, were incurred, or otherwise took place.

The emphasis is on factual—the person preparing the report must show as factual in the report only that information which he/she objectively includes from such source documents or transcripts as are purported or deemed to contain information reportable as factual.

For example, if the business writer uses the month-ending inventory computer report that was provided to him/her by the General Accounting Manager, and the business writer reports from this source to the Financial Vice President the "Inventory Status Report," then it is a factual and informational business report. However, if the General Accounting Manager provides these amounts in a memo only to the business writer, then the writer has the obligation to report the amounts, but state in a cover note or in a footnote the source of this assumed factual information.

Further, if the business writer takes the same exact computer data and reports it to the Financial Vice President, but with his/her analysis of "Better/Worse than Plan" results and explanatory causes, then this is an analytical report and not an informational report.

Informal reports range from the Purchasing Agent's report to the Engineering Department about a calibrating measuring device to the Controller's report to the President about the operating income of the divisions and subsidiaries. The Controller needs to write the report to the President much differently than the Purchasing Agent does to the Engineering Department.

Moreover, not all reports deal with monetary information. When the interviewer from the Personnel (Human Resources) Department reports about his/her test results from an interview of a job applicant to the Sales Manager, it is also a report, although there may not be any monetary amount mentioned at all. Nevertheless, this is an informal report.

What is most important to the business writer whenever he/she writes any report or issues any written communication that appears as a report of facts, data, information, etc., is that it must be factual, i.e., based on facts, data, and information that has been obtained by or provided to the business writer.

### 6.4.3 Formal Reports

This section will focus on presenting the format and parts of the formal report. The following is an all-encompassing description of these in a most complete, full-blown formal report. Such complete formal reports are seldom warranted by most business writers in companies. Usually, most of the lengthy formal reports are prepared without several of the parts shown below.

For the business writer to elect to prepare such a complete formal report, the subject matter, issues, and problems in the particular business situation should be of sufficient importance to the well-being of the company and should be sufficiently complex—otherwise, he/she should select the informal business memorandum as the vehicle to write about the matter. For example, the feasibility study for establishing a new factory in Arizona or a subsidiary in Thailand warrants a complete formal report. However, the business situation in the import licensing of machinery example presented in Chapter 5 would require only a thorough business memorandum, albeit structured in parts by headings, similar in this respect to the content (body) of the formal report itself.

A complete, full-blown formal report has the following structure and parts, and is affixed into a cover or binder:

- Half-Title Page (Title Fly).

  This page should include the title, name of the author (business writer), and date of the report. Often, the name of the person or business unit (such as CEO's Office, Asia/Pacific Group) who requested the report, or to whom the report is directed, is shown on this page.

- Authorization Document

  When the report is prepared in response to a request by memorandum, letter, meeting agenda, or other document, this request document should be included in the authorization section. It is recommended here that the business writer overlay this document with a page stating that the report was prepared in response to the attached authorization document. This is effective because the recipients then instantly see that someone other than the writer also considered the topic of major importance for a formal report.

- Transmittal Memorandum (Letter)

  As an alternative to the authorization document, or in addition to it, the business writer may decide to prepare a memorandum (internal) or letter (external) to address the report to the recipient and briefly state the subject matter, conclusion, recommendation or advisory, or the highlights (informational report) from the report. This sometimes takes on the appearance of a preface.

- Abstract, Summary, and the Executive Summary

Even in the case of the formal business report, very few are so complex or important that a separate abstract (or synopsis) or summary should be prepared and placed in the preliminary part of the report. (In the case of scientific or technical reports, abstracts or summaries are customary.)

Conversely, it is customary in the case of not only formal reports, but also of the lengthier (seven pages at least) informal reports—and even with lengthy memoranda—to prepare an executive summary. Below is a brief description of these:

*Abstract.* While an abstract is like a brief summary, it is different because it should only indicate what is in the report. It is a somewhat detailed description of the table of contents or list of contents that should not exceed a typewritten page, or 200 words.

*Summary.* While the summary is like an abstract, it is longer (it can be two pages for a 20-page report), and it provides specific information from the report or a description of its content. Also, the summary should state sufficiently the conclusions, recommendations, or advisories from the report which is absent in the abstract. In fact, the summary is a scaled-down version of the complete report. (An author used the term "epitome" in place of "summary" as the title for the summary part of the report.)

In some books on business communication, abstracts are referred to as "indicative" and summaries as "informative" types of summaries (the terms "synopses" and "abstracts" are used interchangeably, with summary being the most common inclusive term).

*Executive Summary.* Always capitalized in its title, the executive summary should show all the main or critical elements of the report in a bullet-like format. It is a birds-eye view, an overview, of the whole report for a quick understanding. It must always include the summary of the conclusions, recommendations, or advisories. It should be written in crisp, effective, attention-grabbing topical sentences such as: "Import permissions appear discretionary by Thai Government: Detailed investigation required as outlined in II. A. 3." It is a separate page, preceding the text of the report itself (but it is after the table of contents page).

- Table (List) of Contents

  If the report is long enough, then a table or list of contents page is necessary. This should be a listing of the captions or headings present in each part of the report, along with their respective page numbers. The appendix, bibliography (rare in reports), list of the exhibits, charts, tables, and/or figures are also listed.

- Content (Report Proper, Report, Text, Report Content)

  In several books on business communication, the authors do not delineate at this point in the discussion "content" or, better yet, "report content." They discuss the "introduction" section of the report after the table of contents and go on all the way through "support material" in piecemeal discussion.

  Here, the main parts of the formal report are listed, because some of these parts are ineffectual, if not erroneously, mixed (commingled) by business writers, as discussed in more detail later on in this section. The main parts of the content of the formal report are listed as:

  - Introduction
  - Body (further delineated by the headings in the body of the report)
  - Ending Summary
  - Conclusions
  - Recommendations (Advisories)
  - Addenda

  But, in analytical reports, where several governmental, business, academic, periodical, and/or other source material or documents have been used as "research" in the report, or as addenda to it, the business writer should consider adding a secondary summary, such as: "Summary of Material and Documents."

The above list of main parts is for "inductive arrangement reports." In case of "deductive arrangement reports," it is important to include another "conclusions" part right after the "introduction" part. In this advance "conclusions" part, the writer presents what conclusions he/she expects the recipients to draw or derive after reading the report, and the reasoning for such conclusions.

Following is a brief discussion limited to the essential principles of each part of the formal report.

- In most books on business writing it is common to instruct the reader that the *introduction* part of the long report should provide its readers with an orientation about: the topic (subject matter), why the report was written about the topic (purpose), the scope of the report, the source of material and information, and a preview of the report (its parts, order, and main ideas—a guide on how to read and understand the report).

  Here, we also view it as essential for the business writer to consider carefully—to visualize—the background and probable comprehension level of the recipients of the report's specific topic and content. The business writer has his/her best opportunity in the introduction part of the report to provide a sufficient understanding to all recipients about the report, even to those who would not understand some or many of the subject-specific terminology, concepts, or details.

- The report's purpose—the "heart of the matter"—is to present to the recipients the information, explanation, analysis, problems, and associated factors as these relate to the subject matter of the report. This part of the report is usually called the *report body (body)*. It is the longest part of most reports; indeed, it is the report per se.

  The body of the report should be subdivided by carefully selected topical headings to group the elements of the report into meaningful separate sections. For example, in a feasibility study report of the new subsidiary in Thailand, a meaningful heading would be LABOR LAWS AND LABOR COSTS IN THAILAND.

Under such headings, only the text of the report should be included, with references to the supporting attachments or addenda, such as documents, charts, exhibits, etc. That is, the text under the headings should not be interrupted by presenting such supporting material within the body of the report, except perhaps for very short items, such as a formula or calculation (for example, for ROI).

Reports vary to such a degree that it would be misleading here to advise the business writer of the "right way" to organize the body of the report. The important point is that the business writer consider the unity, logical flow and interrelation (association) of points, and the chain toward the conclusions of the report (although the conclusions might have been shown as prefactory already).

- The *ending summary* should be very brief, showing only the major findings or highlights. (In contrast, in the abstract, synopsis or summary in the prefactory part of the report—if such is prepared—the review of the report should be much more detailed.)
- In this end part of the report, the business writer provides his/her *conclusion* based upon the purpose, facts, supporting material, and analysis/explanations provided by the writer in or attached to the report. It is essential that the business writer does not suddenly—as if in an afterthought—invoke any new ground for the conclusions that he/she did not sufficiently present in the report itself. This would be a fallacy. Hasty conclusions also should be avoided by the writer.

These are the business writer's own conclusions, even if no other conclusions can be drawn by the prudent recipient. Therefore, the business writer should not use phrases such as "Undoubtedly we can conclude...," "It is obvious that...," or "No other conclusions can be drawn but...." Not only is this a bad tone, but it creates problems for the writer when one or more of the recipients replies with different conclusions drawn from the report—or for enlargement or expansion from the report based on additional data of which the writer was unaware or omitted.

The conclusions part may at times be a couple of pages long, depending upon the complexities of the report and the problems contained therein.

- Based on the report and his/her conclusions, the business writer most often does offer his/her *recommendation* for various action, decision, further and expanded evaluation, abandonment of the project, or various other recommendations. Even in the case of a lengthy formal information (informative) report, it is essential for the business writer to provide his/her conclusions, and especially recommendations—otherwise, the writer should have written a memorandum and not a structured formal report.

  This is where the writer has to be especially careful, because others may act on his/her recommendations. In fact, they usually do, because they deem the writer of the report an expert on the subject matter, since he/she was selected or requested by one higher up in the organization to write the report.

  The careful business writer should hedge the recommendations, and/or state the uncertainties, unknowns, potential deviations (glitches) in the eventualities or in the subject matter. Including recommendation alternatives A, B, and C is prudent approach if the situation so warrants.

- The term *addenda* refers to all the support material that is attached to the report. Usually it consists of the appendixes—the documents, schedules, charts, and other supporting materials. These materials should be clearly marked or labeled (such as Exhibit A, B, etc.) for reference in the report's body, and listed in the table of contents.

  Very rarely does any report require the use of "works cited" or "bibliography" or other "reference" listing. If present, however, it is included in the addenda part of the report. Even rarer is the use of an index in a report.

### 6.4.4 Pointers for Report Writing

The following pointers are unique to the writing of long formal reports:

- Structural Coherence

  In Section 6.4.3, the parts of the long formal report are presented in their conventionally accepted sequence, delineating introduction and conclusions (as prefatory and ending parts).

  However, in a well-composed and organized report, in each heading part of the body the writer should use a network (interconnection) of introductions, explanations, discussions, and conclusions, as if walking the readers by the hand through the report. (It is easy to grasp that under a heading of LABOR LAWS AND COSTS IN THAILAND a separate set and chain of introduction to conclusions is essential.)

- Comparative Reports

  The term "comparative report" is unique to this book, but it applies to a variety of long reports. The reason for isolating the following type of reports into the category of "comparative reports" and discussing them in this separate section is because such reports pose a major organizational dilemma for the business writer, unlike other long reports.

  Let's select for discussion purposes three business situations as subject matters for formal reports assigned to the business writer:

  - Evaluation of three different drug products, A, B, and C, for market penetration into Country XYZ.
  - Evaluation of three different countries, A, B, and C, for establishing a new manufacturing subsidiary of the corporation.

- Evaluation of three different computers, A, B, and C, for implementation throughout the corporation.

It is easy for the reader of this book to realize that in each of these reports the business writer will have several critical factors that he/she has to separate into subparts (headings) for topical information, explanation, analysis, evaluation, conclusion, and recommendation.

There are many reports written in the business world which comparisons of two or more products, divisions (A vs. B), companies (say suppliers or law firms), regions or countries, etc. are compared by the business writer, not only for selection but also for performance or other evaluation.

## 6.5 Proposals

One of the most specific types of business writing is the "proposal," yet it is the least defined type of writing in most books on business writing. Usually, proposals are described as sales tools, or as offers to supply products or provide services. This is true in the sense that proposals are offers for the consideration of the recipient.

The most descriptive definition for the term "proposal" is its legal definition: "An offer by one person to another, of *terms and conditions* with reference to some work or undertaking or transfer of property, the *acceptance* whereof will make a contract between them."

However, in the business world the term "proposal" is used more restrictively, usually only for the type of detailed written business memorandum, letter, or report which encompasses the elements of the proposal expected in business.

Such proposals are of such magnitude and diversity for the company or organization that the proposal is invariably prepared by a "team," of which the business writer is only one of its members, albeit for the purpose of this book he/she is the member who "puts together" the proposal.

Proposals are prepared either in response to formal requests or bids (governmental requests), that is, they are "solicited" by someone, or "unsolicited," that is originated by the company or organization to offer (solicit) its performance, service, or product to another company or organization for monetary or other compensation (sometimes the proposal is made to the government). When the proposal is solicited, it is often referred to as Request for Proposal (RFP), usually only as RFP.

### 6.5.1 Forms and Content of the Proposal

In the case of RFPs—especially from the government—the format for the proposal is usually prescribed in detail.

In other cases, the usual proposal format is structured as follows. It is very similar in most respects to the long formal report; therefore, only the aspects which differentiate the proposal from the long formal report are discussed here.

- Preliminary Parts

  For the lengthy proposal, the same guidance applies as for the long formal report in Section 6.3—above: half-title page, title page, authorization, transmittal, table of contents, abstract, summary, and executive summary.

  The business writer assigned to put together the proposal should carefully evaluate how the abstract and the summary need to provide the essentials of the whole proposal. The writer should discuss with, or obtain from, the various proposal team members their own suggested texts for these parts, and then the writer should reduce these into a text of manageable length by taking the essentials from each (i.e., from the Controller's, Engineering's, Production's, Marketing's, etc. sectors).

  These parts are prepared after the report body (here the proposal body) has been completed and agreed upon or approved in its entirety.

- The Proposal (Body)

- Introduction (Background)

  In the opening section, the problem or task for which the proposal is offered by the company should be well described. Even when the proposal is a "RFP," that is, the problem or task has been provided in detail by the other party (private company or government), it behooves the business writer to recite and enlarge upon it by incorporating the proposal team's expert grasp of the situation. This will show that the business writer's company has a handle on the problem for its solution.

  As with the long formal report, it is prudent to add a conclusions with recommendations part after this project description section, a sort of deductive arrangement of the proposal.

  The business writer should bear in mind that whereas many of the long formal reports are only read, the proposal is *evaluated* in each of its parts and on the whole by the recipients for *acceptance.* Therefore, these leading sections are critically important to persuade the readers of the proposal that the writer's company "knows what it is talking about," understands the situation, and has the solution to the problem.

- Summary (Preview) of the Proposal Body

  The same method is applicable as was described in Section 6.3 for the long formal report.

- Body

  The proposal is unique among business writing for the following reasons:

- As in the long formal report, the various issues must be sufficiently presented, discussed, and explained before such conclusions are reached.

- However, the proposal has to include the unique features of *plans* (see Section 6.6), presenting how the writer's company plans to approach and solve the problem, or what the company plans to perform for the recipient company or organization.

- Thus, whereas in the long formal report the recommendations are for the recipients to act upon as the report recommends, the proposal is the writer's company's offer to proceed along its plan shown in the proposal to carry out the project or solve the problem on behalf and/or in cooperation (as is usually the case) with the recipient company.

Each proposal has its own structure, depending upon the subject matter. Often, especially when it is a RFP from the government or from a large company, the structure is dictated by the detailed written material provided by the proposal-requesting party.

The usual main parts of the proposal body are arranged along the lines of:

- Technical Plan: "Technical" means production, financial system, feasibility study, supply of products (such as in governmental RFPs), or whatever the subject matter is. It is the technical plan as proposed by the writer's company to the recipient party. Here, the knowledge, comprehension, and grasp of the technical issues need to be illuminated by the company.
- Management Plan: This is the company's outline of how it plans to carry out the task with its own resources and personnel. In this part, the company usually includes the description of the expertise of its personnel who will work on the project—it is the demonstration (the "selling") of the company's capabilities.
- Resource and/or Assistance Requirements: Often overlooked in other books, this section is prudent to include in detail as a separate part, or as subparts, in the other main sections of the proposal body as to what will be required in terms of investment, cost, cooperation, and/or personnel assistance from the recipient party.
- Compensation, Fees and/or Costs: The primary element here is the compensation or fees—sometimes called the costs—that the proposing company will expect to receive from the

recipient party. This needs to be spelled out in every detail, going down to the nitty-gritty "out-of-pocket expenses," because the contract should be drawn up in similar detail. Moreover, these factors should be presented as tied-in to conditions and assumptions about scope, time, predictable or unforeseen lesser/greater workload, etc.

- Conclusions: Usually the proposal does not require a conclusions section, but it might be prudent at times to prepare it to highlight important points from the proposal.
- Addenda: Attach whatever supporting material is referenced in the proposal or is helpful to win the approval for the proposal. In addition to the discussion about addenda in Section 6.3, the company should include in the proposal addenda one or more examples or even copies of news clippings about its relevant "success stories," and the background profiles of the key company personnel who will carry out the project.

### 6.5.2 Language of the Proposal

The proposal should be written in language that is objective, toned down (unemotional), and persuasive; the "positives" of why the company should be approved for the proposed project should be emphasized.

It is essential to write the proposal in the clearest manner possible for comprehension and understanding, as well as for favorable reception purposes. The recipients will have enough of a burden with the technical part (plan) of the proposal—therefore, it is helpful if the company alleviates cumbersome or faulty English. Everything relevant should be explained, leaving no room for misunderstanding.

In the case of a complex or highly technical proposal, it is an effective tool to attach to it a cover letter with a *suggestion* for the recipients to request clarification or explanation to avoid misunderstanding.

Oversimplification, undue optimism, bragging, and comments about competitors should be avoided in the proposal.

It is a sign of professionalism and honesty to state in the proposal that given problems are not yet understood and how the company plans to analyze or investigate these in the course of the project to develop solutions for them.

## 6.6 Plans

Corporations, organizations, and governments have long-term existence. They expect this to continue—they operate for the long-term future, as well as for the monetary present. The past is of importance to such organizations; however, the years or decades to come are of far more importance to executives and managers who must plan for the future welfare of the organization.

In every company and organization, myriads of plans are formulated. A *plan* is the establishment of *goals and objectives* and the determination of the optimum ways to accomplish what was *planned.*

Planning, i.e., the activity of the formulation and preparation of plans, is a primary function in business (as well as in government or any organization); all other activities and functions depend on plans.

A plan of action, even in the most instant situation, is still a plan that requires at least instant planning even if it is only in the mind of the individual.

The plan is differentiated fundamentally from other written business communications, because it sets the tasks that must be accomplished at some future date, ranging from next week to years to come, instead of at the present time.

The business writer who is involved in the task of preparing or writing about (say analyzing) any type of a plan, needs to understand conceptually the essentials of the plan, the planning process, and its criteria. Particularly, the writer needs to understand the differentials of the mental process, as well as the differentials in the written material for past, present, and future business activities and situations.

### 6.6.1 Types of Plans

It would require a sizable book just to cover sufficiently the fundamentals of preparing the profit plan of a corporation, which is the usual topic of books written on planning.

The feasibility plan, the annual profit plan, the long-range (strategic) plan and such subject-specific plans as research, marketing, production, import, etc. can be covered adequately only in separate books on each.

As far as the business writer is concerned, for the task of his/her business writing, plans range from next week's production plan to the five-year plan—the term "plan" is applicable to both. That is, the business writer is involved with the process of planning in both, albeit on vastly different levels and degrees.

### 6.6.2 The Business Writer and the Plan

The most basic characteristic of the plan is that it involves the mental process of foreseeing and forecasting future conditions, situations, and circumstances by hypothesis, assumption, analogy (to know events and facts), and other elements.

The plan of the tasks must be established to be valid and correct, not in the past or present, which are comparatively known or can be ascertained, but in the assumed business environment of the future.

However, when writing a plan, the business writer must first prepare a detailed written outline of what the future "picture" looks like to him/her in composite of all assumed involved factors. Moreover, at this early stage, the business writer must expand this picture by writing down the potential foreseeable alternative adverse variables, in order for him/her to plan to prevent or cope with such contingencies.

These require the business writer to write the plan in a language and in terms that ensure that the recipients and readers of the plan know what the writer is talking about in terms of assumptions, forecasts, and hypotheses versus the actual and factual data, and information components included in the plan.

Here, the discussion of the plan and the process of planning is intentionally limited, because it cannot be properly discussed and covered in a fragmentary and isolated manner, not even by selecting a production plan, a departmental budget, or a Product XYZ marketing plan. This would mislead the reader into following such partial guidance and not knowing what else are essential components and elements of the plan writing process.

There are several excellent sources which provide professional guidance on various types of plans and planning, considerably beyond the scope of the books used in business schools.

## 6.7 Presentations

Often, it is required that the business writer prepare and submit in advance of a meeting or conference (internal or external) a written presentation of what he/she or other members of the company will present or discuss about the subject matter.

Presentations are important vehicles of business communication—meetings and conferences are held about matters considered to be important by the attendees.

Presupposing that the writer knows why (purpose) and what he/she is to write about, the business writer first should prepare the written presentation with a focus on the clarity, correctness, accuracy, and soundness of its technical part. For example, in preparing a presentation on the topic of zero-base budgeting for an internal meeting, the primary focus should be to ensure that the written presentation is technically correct in presenting all the essentials of this distinct budgetary concept and method.

Another example is preparing a presentation on the past performance and future marketing plan for a given product. The business writer should focus at this stage on presenting accurate facts, data, forecasts, analyses, and surveys which are factual and supported by documentation in the appendix.

The next stage in the process of preparing the presentation is for the business writer to determine the *specific purpose* of the presentation and incorporate the specific purpose into the written presentation.

In the above examples, it may be that the business writer wants or needs (for example, if the writer was requested by the president of the company to do so) to present zero-base budgeting as undesirable to the company, while the business writer presents the given product for the purpose of concentration of marketing efforts on it throughout the company.

When writing the presentation, the writer should *consider the audience* (recipients) of the written presentation. Here, it needs to be strenuously emphasized that the business writer should write the presentation most *objectively*. It will be subsequently orally presented (as written or based on it) by the writer or someone else to an audience, and it should not be tilted or favored towards any of the planned attendees, regardless of their level of position.

At this stage, the business writer is able to adjust and expand the written presentation for the recommendations, persuasiveness, or inducement intended by the writer or whoever assigned its preparation. The important point is that the subject matter itself should be presented in an objective manner, separated from the recommendations or persuasive/inducible comments.

Whereas a long formal report may be well over 200 pages plus the addenda which can be read (and reread) by the recipients at their own pace, the presentation itself has to be limited to a length that can be orally presented in no more than 30 minutes of speaking time. (With pauses, emphasis, eye contact for effectiveness, etc., this is about 45 minutes in actual time.)

The written presentation is similar to and has many of the same essential elements as the long formal report, including with regard to its format and packaging.

## 6.8 Procedures, Handbooks, and Manuals

Procedures (standards) in any sizable corporation constitute several different handbooks and manuals. Even within such a specific business discipline as financial management in corporations, there are separate controller's manuals, treasurer's manuals, auditing manuals, planning manuals, tax manuals, and perhaps even other separate manuals.

The titles "handbook" and "manual" are often used interchangeably; however, usually they serve different purposes.

Manuals are comprised of individual procedures, and the procedures serve dual purposes:

- First, procedures set forth the rules, regulations, and policies of the company for its employees.
- Second, procedures describe for the employees what and how they must perform according to these rules and regulations, and what they must do to carry these out properly.

While it may be best to write the procedures in two versions or in two parts, keeping the above separation in mind, this can seldom be accomplished; nor is it practical to do so, since the two purposes are interrelated.

Therefore, the business writer should place the emphasis on providing detailed instructions, while ensuring that the rules and regulations are accurately, fully, and specifically covered in the procedures.

The business writer should aim at writing the procedures with their audience in mind, since they are the persons who must adhere to them. For example, in a procedure on monthly capital expenditure reporting, the emphasis must be on who, what, how, when, and where to report such expenditures, providing the specific instructions, including how to tabulate the data. It is all too easy to elevate the topic, such as including in this example more than a necessary discussion of the importance and managerial use of the reported data.

When writing procedures, the business writer is restricted in tone by the requirements of using the ordering, commanding words of "shall," "will," and "must."

Many procedures are substantially "overwritten" and are far too long. The business writer should aim at writing the procedures concisely, editing out unnecessary verbiage.

The one major advantage to the business writer who writes procedures, as compared to letters, memoranda, or reports, is that he/she has many other already published procedures to follow as examples. Indeed, the writer must adhere to the company's method of preparing procedures. Nonetheless, each procedure is unique to itself, and the business writer must develop it as individual writing.

## 6.9 Abstracts and Reviews

At times, the business writer may be requested to prepare for executives, a committee, the CEO's or CFO's office, or for a meeting (or conference) an abstract of a long memorandum, letter, report, proposal, or even article or book (the latter if such is pertinent to the company or if it contains a new concept, method, etc. applicable to the company).

In such a case, the business writer needs to reduce the substance of the material but ensure that the abstract does not distort or change its meaning, nor its emphasis. The abstract should state exactly what the author of the source writing presented or said. The business writer should go as far as "copy" the first person usage, e.g., when in a letter: "I have identified..." should not appear in the abstract as "The author (or Jack Brown) identified..."

The business writer of the abstract should make sure that the recipients of the abstract understand that he/she merely faithfully abstracted the original writing for them, as he/she determined what are the essentials for the abstract.

At other times, the business writer may be requested to prepare a review, usually of an article or book. While this also should be prepared faithfully with regard to the content, here the business writer is expected to provide his/her evaluation and opinions to the recipient.

Sometimes, "analyses" is covered as a distinct type of business writing, for example, an analysis of a report. In this book, this is considered part of the memorandum or letter, or occasionally as a report (even when it is about another report).

## 6.10 Responses to Communications Received

Often, our business writer is in the first place the recipient of a business letter, memorandum, or report, and he/she is required to respond to it in writing.

It is most important that when the roles are reversed, and the business writer becomes the recipient of written communication, he/she reads, studies, analyzes, and evaluates the received business communication with the same diligence and care that he/she utilizes in his/her own writing. The experienced business writer possesses comprehension skills that enable him/her to grasp all the material and nuances present in others' writing.

Whatever is the nature of his/her reaction, it should be in response to the received written communication. That is, if the business writer goes beyond the scope of the received material, or introduces extras (be that information, concepts, other) in the response, then he/she must state, identify, and explain these—otherwise, it is not a pure response but a new writing.

# CHAPTER 7

# Writing Style

## 7.1 Format of Written Communications

Format is integrated under the various types of written business communication as to their integral elements, instead of separated for discussion purposes.

## 7.2 Diction

Diction here refers to the business writer's choice of words, especially with regard to clearness, correctness, exactness, and effectiveness.

Without specifically using the term "diction" throughout this book, the concept of using proper word choice has been emphasized. Here, diction is discussed in the context of proper English usage by the business writer.

Basically, the three aspects (qualities) of diction are:

- Appropriateness—selecting the most appropriate words for each sentence in the context of the writing, even if these are "jargon." (*Terminate* the project may be more appropriate than *end,* even if *end* is a more common word.)

- Specificity—using language and words as specific as possible. (*Terminate* is more specific than *end.*)
- Imagery—using words to suggest the desired images, and figures of speech (similes, metaphors, analogies, allusions, other).

### 7.2.1 Abstract Words

Improperly used or carelessly chosen words are often the problem in unclear communication. Some words are abstract, too general, or have hazy meanings.

"Rapidly," "promptly," or "as soon as possible" are too vague and do not provide a precise understanding of the timing factor involved. Even the specific term "winding machine" can be too general, particularly when the "machine" is not a complete unit without its accessories.

"To obtain" can cover a wide range of activities in the business world.

**Exhibit 7.2.1**

**JARGON: THE LANGUAGE OF BUSINESS**

**THE ABC LIST**

Accelerated
Accomplish
Accounts Receivable
Accrue(d)
Activity
Adversarial
Advocate
Affiliation
Aging of Payables
Allowances
Amortization
Analytical
Articulation

Assets
Assumption

Bad Debts
Balance (Balancing)
Base points
Book value
Bottom-line
Break-even
Breakthrough
Bullet

Capitalization
Carrying
Cash flow
Centralization
Commerce
Commensurate
Consistency
Consummate
Conversion

### 7.2.2 Jargon (Shoptalk, Technical Terminology)

Nearly every book on writing, especially on business communications and writing, provides the uniform advice that the writer should avoid using "jargon" in his/her writing. Most of these books do state, however, that when jargon is used as "shoptalk" of the profession or trade, it is not only permissible but is required as "learned" language, for its use demonstrates professionalism on the part of the business writer.

This appears contradictory, mainly because of the absence of a detailed, clear definition of the several meanings of the word "jargon" in these books. The definitions of jargon may be listed in summary as:

- Originally, jargon meant meaningless, foolish, nonsensical chatter, babble, or prattle, which could be best grasped by the saying "talking off the top of one's head." (Jargon as meaningless chatter; chatter being synonymous with babble.)

- According to dictionaries, jargon has come to mean language vague in meaning, inappropriate, full of high-sounding words and circumlocutions (verbal evasions, wordiness). Such jargon should be avoided by any writer, but foremost by the business writer, as advised throughout this book.

- Jargon also means the specialized (technical) language of given professions, trades, or other groups, which is not readily understood by the uninitiated person. (By technical, it is meant to refer to financial terminology, or engineering terminology, for example.) For a detailed discussion, see Chapter 4 of this book.

The last definition of the word jargon is "learned" and professional, provided that it is either used among people who know or should know such jargon, or that the writer explains the meaning of the jargon in clear and precise terms, aimed at the level of comprehension of the recipient.

The invalidity, or rather inapplicability, of advice in most other books about the use of jargon arises from the authors' lack of familiarity and experience with what constitutes the technical terminology (jargon) of the individual business professions, or even of business in everyday usage. For example, in all books the authors would advise the professional usage of jargon, such as the words *inventory, assets,* and *fliers* (in advertisement), because these words are accounting or marketing terms that have no common English language counterpart.

Most authors do not realize that there is a myriad of "learned" words that are not purely technical (e.g., *inventory),* which became just as much a part of the business lexicon in usage—despite the harangues against their usage in most books on business communications. They consider such words abstract, showing a fondness for "learned" rather than ordinary (unpretentious) language.

The following is a list of common words that other books denote as falling into the category of jargon. These are but a handful of the words listed in books on business communications as being pretentious, inflated, and used to show off knowledge of big words when little or common words would do just as well. The substitutes recommended by the authors for the reasons cited above are listed beside the words:

| | |
|---|---|
| accomplish = do | prioritize = rank |
| approximately = about | remittance = check, payment |
| execute = sign | terminate = end |
| facilitate = help | utilize = use |

However, these authors do not realize how widespread and entrenched (that is, common) the use of such words became in the business world. Nor do they realize that their offered substitute words do not mean the same in business parlance as the words they deem as undesirable jargon.

For example, to *execute* an agreement or a contract encompasses the *signing,* but the signing is only a perfunctory (albeit critical) element of the process of executing it. In business, *approximately* has the quality of being acceptably close (even "closely approximate")—like the values or prices of two different products—whereas *about* is used in business, as in common language, as a vague and general term.

Moreover, if the business professional who writes letters, memos, or reports on behalf of the executive(s), or representing the company, department, or subsidiary would follow the advice found in other books and substitute the recommended words for such "jargon" words, it would be just a matter of time before such serious correspondence is reassigned to others for writing.

Many persons have carelessly misused jargon and this has cast discredit upon its use.

A type of jargon has been so overused in letter writing that readers of correspondence view such writing as merely trite phrases, and adversely view the writer of the material as insincere and shallow. Examples of this type of jargon are: "We thank you for your interest in seeking a career opportunity with... (but regret to advise you)...," and "We are most pleased to take this opportunity to inform you that..." These are examples of pretentious use of higher sounding language than is commonly used in English. In normal, everyday language, the latter expression would read as: "We decided that... (and state the information)."

## Exhibit 7.2.2

## FREQUENTLY MISUSED WORDS IN BUSINESS WRITING

**Among/between.** Among is used for three or more people or objects, while between is used for two.

> Distribute the earnings among the stockholders.
>
> Divide the workload between Personnel and Accounting.

**Amount/number.** Amount is generally used to refer to things or a monetary sum that cannot be represented in an exact figure, while number refers to things or amounts that can be counted.

> The amount of time that would have had to been expended made the project unfeasible.
>
> Due to downsizing, the number of our employees has decreased significantly.

**Balance/remainder.** Balance is an accounting term that refers to the difference between the debit and credit sides of a ledger account, while remainder refers to something left over.

> Once the cash balance of each account is determined, report the information to corporate.
>
> The remainder of the material should be sold as scrap.

**Correspond to/correspond with.** Correspond to is used when one thing is similar to another, while correspond with means to exchange letters.

> The examples you utilize do not correspond to each other.
>
> I have corresponded with my counterpart at the division for years.

**Disinterested/uninterested.** Disinterested conveys the quality of neutrality, while uninterested conveys lack of interest.

I chose the course of arbitration, because I wanted the matter decided by a disinterested party.

The manager is uninterested in discussing matters that don't relate to the operation of the plant.

**Formally/formerly.** Formally is used to discuss something done ceremoniously or done according to an established method, while formerly refers to something that has occurred previously.

Only formally prepared requests can be sent to the Minister of Finance.

Formerly, such practices would have been viewed as being unethical.

**Forward/send.** Forward is used when an object is being sent ahead; while send is used when a means of communication is dispatched by the individual who initiated it.

Mr. Clark is no longer with our firm, but we will forward any correspondence to him that is sent here.

Please send me any information you have in your files.

**Infer/imply.** Infer means to conclude as a result of facts; while imply means to suggest indirectly. Readers infer, while writers imply.

I infer from the President's memo that further cut-backs are imminent.

Do you mean to imply that the company functioned more smoothly under the old guidelines?

**Majority.** Do not use the term majority when referring to a singular object.

INCORRECT: The majority of the report had to be rewritten.

CORRECT: The majority of the sections in the report had to be rewritten.

**Respectfully/respectively.** Respectfully means with deference; while respectively means in the order given.

I respectfully ask Your Honor to reconsider.

The highest rates of plant productivity were determined to be in Taiwan, Sri Lanka, and Thailand, respectively.

### 7.2.3 Weak Verbs

The business writer should avoid excessive use of linking verbs such as *be, seem, appear, look, feel,* or *become.* Over use of these verbs makes the written material bland and monotonous. However, use of linking verbs is effective when indicating logical equivalents. For example: Most of our legal team is in Chicago.

Use of verbs such as *occur, prevail, happen,* and *exist* also create undynamic writing. Try to choose sharp, vivid verbs to describe the action. For example, in the two sentences below, the latter is more direct.

The drop in our stock prices occurred yesterday.

Our stock prices dropped yesterday.

### 7.2.4 Clichés

Clichés are overused phrases, used thoughtlessly by writers. They are often merely trite metaphors and similes, and the use of such phrases demonstrates unoriginal thought on the part of the writer. Such phrases, more acceptable when spoken, should be avoided by the business writer.

Commonly used clichés include the following:

busy as a bee
doomed to disappointment
get the show on the road
gone off the track
really down to earth
uphill battle
worth its weight in gold

Moreover, most of the clichés are vague and non-specific. It is non-definitive how busy a person is if he/she is "busy as a bee."

Some sources include terms like *bottom-line* among the list of clichés; however, the business writer must be careful to recognize the difference between clichés and jargon; the latter is acceptable if used properly. ("Bottom-line" has a very clear meaning to accounting, finance, and management personnel.)

It is not acceptable to include clichés enclosed in quotation marks, such as: The feasibility study really "hit the nail on the head." This practice tends to draw even more attention to the cliché.

### 7.2.5 Mixed Figures of Speech

Mixed figures of speech are confused and bizarre expressions, employed by writers who do not think logically about the text that they write. For example: Unless the problem is resolved we'll end up paddling upstream behind the eightball.

Figurative language (such as "behind the eightball") should, in any event, be used only in very rare cases by the business writer.

### 7.2.6 Empty Intensifiers

While in conversation, intensifiers, such as *very, so,* and *much,* are often used to give emphasis, these words weaken the message in the written form.

Business writers should write as simply as possible, however, use of empty intensifiers suggests the writer's lack of diversified vocabulary.

> WEAK: I'm really happy to hear about your promotion.
> BETTER: I'm delighted to hear about your promotion.

Often, intensifiers can be omitted altogether.

### 7.2.7 Vocabulary

Vocabulary is defined as all the words found in a language. Everyone possesses a given level of vocabulary.

Businesspersons, especially those in positions that involve them as writers or recipients of business writing, accumulate an extensive vocabulary. Through formal and/or informal education, or even by just studying the basic business and management books that are essential to learn the fundamentals of the various business files these individuals have to acquire an extensive vocabulary.

Aside from providing, for finance, production, procurement, regulatory and other management areas, the language peculiar to these fields, the writers of these books use a rich vocabulary that the readers acquire, either consciously or subconsciously.

Businesspersons are usually avid readers of newspapers, periodicals, books, and volumes of written communication, including their companies own mandatory procedures, manuals, and directives.

Therefore, the business writer should write in a language that is at the level of the recipients. Although avoiding the use of pompous or formal language and vocabulary, the writer should judiciously use words that express the precise meaning he/she desires to plant in the minds of the recipients.

A few examples will illustrate this approach.

- Approximate. This is a term frequently used in business to describe amounts or positions on issues as being near to another, more or less being correct or exact. The word *about* does not provide the same close meaning when writing in business.

  In governmental regulatory language, which aims at providing precise directions to business, even the term "closely approximating" is used, with the criterion identified as what it means (say plus/minus five percent deviation from a standard).

- Disincentive. This is a term used in business for factors that should deter action, plans, or policies from taking place, being formulated, or being implemented. "Disincentive" is a more encompassing term than "deterrent."

  A deterrent implies, connotes, or describes that the action, plan, or policy should not take place because of legal, regulatory, or other potential or real repercussions.

Further, the word *penalty* is not a substitute for disincentive, albeit the writer may determine that *penalty* is the correct word to use, because penalties will be incurred if the plans or policies are implemented.

- Expedite. In business this term is used particularly to obtain or deliver products or equipment in a businesslike, orderly fashion, taking into consideration the various factors involved. It is a learned business word, and the process of expediting is more involved than the common term "speeding it up."
- Remuneration. This term often refers to "pay," but it is usually used for a package of monetary and other benefits or compensation to employees. The business writer should use the term "remuneration" when it is applicable, such as in the human resources context.

On the other hand, the business writer should be careful to avoid misusing words, which reflect negatively upon the writer and his/her department or the company. Moreover, misused words can convey the wrong meaning in written communication.

Exhibit 7.2.2 lists a few examples of words that are frequently misused in business and even in media communications. The words selected for this exhibit are misused for other reasons than being homonyms.

## 7.3 Sexist, Ethnic, and Racial Language

Language is constantly changing. One of the areas where changes have taken place, especially relating to the culture of the United States, is in the attention paid in business, education, government, and media to avoid using *discriminatory* language. Moreover, there are numerous laws and regulations against the use of such language, which is important for the business writer to bear in mind.

Here, we are concerned only with providing the writer with a few pointers about avoiding sexist, as well as ethnic and racial language. These are important essentials of business writing, due to the psychological, interpersonal, and connotative meanings of discriminatory language, which can result in administrative and even legal consequences against the business writer or his/her company.

### 7.3.1 Sexist Language

The noun *man*, as well as the pronouns *he, his,* and *him,* are often used in general terms to refer to an individual. Unless these are used about specific male individuals in the text of writing, the business writer should substitute such words as *one, person, employee,* or *worker.*

When writing about titles or references, it is proper to use *chairperson* instead of *chairman, top people* instead of *top men,* and *staff* instead of *man* (as in "staff the phones").

However, the business writer should be careful not to substitute words that change the meaning of specific terms. For example, while the compound word *man-made* is properly substituted by *synthetic* or *artificial, sales agent* is not a correct substitute for *salesman* because only certain salespersons are agents (employees of the company are not agents). Likewise, *businessman* can be substituted by *business-person* but not by executive or entrepreneur, because not all business-persons are in such positions.

It is an accepted custom in writing today to substitute *he/she* or *he or she* for *he*, and *his/her* or *his or her* for *his.* (This is a preference; in this book the "he/she" version is generally used.)

Another method used by some writers, especially of books, is the switching back-and-forth in the text between "he" and "she," but this can result in connotative meanings or trite usage, particularly when used in examples of business situations. For example, while the Assistant Legal Counsel in the example may well be a "she," in the example of the General Manager of a foreign subsidiary it would be incorrect to use "she" because such general managers are usually men.

Two basic ways to overcome sexist language are:

- Avoiding the masculine pronoun completely. For example, "After the lawyer files the appeal, have him call me" has the same meaning as the gender neutral "Have the lawyer call me after the appeal is filed."

- Using a plural noun. For example, "If there is a question, the Department Head should refer to his copy of the Treasurer's Manual," can be written as, "If there is a question, Department Heads should refer to their copy of the Treasurer's Manual."

Though sexist language is undesirable on psychological and interpersonal grounds, a more serious problem can result if the sexist language is connotative. This holds true with regards to ethnic or racial language as well. For example:

> "In the management audit, the auditors found several serious deviations from the company procedures in her department," instead of in the "human services department." The use of *her* may be construed as connotatively sexist and may result in such charges against the company if "she" (i.e., the Department Head) is demoted or otherwise detrimented.

### 7.3.2 Ethnic Language

The business writer should avoid using words or language that can be construed as ethnic discrimination or bias. The United States has, in the past, fostered several stereotypes related to ethnicity, such as: Latin Americans are hot-tempered, Scots are frugal, Mexicans are lazy, Orientals are inscrutable, and the Polish are not intelligent. These are acquired connotations, and they have implicit meanings.

Business writers do not use such language or slurs. However, the business writer should carefully avoid using words, terms, or descriptions that can result in connotative or implied ethnic slurs or pejorative meanings. For example:

> "The productive volume continues to be below standards, and the output by the *Mexican* workers dropped five percent this month." (If the above refers to the company's Mexican factory, the words "Mexican workers" imply that Mexicans are lazy, not working hard enough. Omitting "Mexican" means the same thing in this sentence—i.e., the output dropped five percent, period.)

### 7.3.3 Racial Language

Unless it is relevant or necessary, reference to race should be avoided by the business writer. For example:

- "The marketing plan for the electronics division was presented by John Smith, a *black* marketing analyst." (Here, the racial reference serves no purpose.)
- If in the above example the sentence continued as "...analyst, and the attendees were left confused," then the word *black* gives a connotative or implied meaning, i.e., that his race had something to do with the lack of effectiveness of the presentation.

Racial reference also should be avoided when it is meant to be positive, such as: "a well-spoken *black* engineer" or "diligent *Asian* workforce."

Not only are sensitivity and proper social behavior at issue, but under various civil rights laws and regulations the business writer or his/her company can encounter problems, even lawsuits, related to racially biased language.

### 7.3.4 Other

Particularly in the United States, though it pertains to most foreign countries as well, bias or discrimination on the basis of religion, age, or disability is also against the law. When writing to a particular foreign country, the writer should ascertain whether the country has a nationally established religion, and alter his/her writing accordingly. Racial and ethnic factors should also be considered.

Thus, the business writer has to be careful to avoid language, data, or words that may result in implied or real discriminatory meaning.

## 7.4 Style of Writing

In the previous chapters of this book, business writing was presented as writing in the business world. This section specifically discusses the "style of writing." For this purpose, business writing here will encompass writing other than technical writing, be it within or outside the company, government agency, or particular profession.

The distinction between technical and business writing is mainly in the subject matter. Technical writing generally deals with the facts of engineering, scientific, or other technical operations. Business writing deals with finance, marketing, and other business situations. Similar techniques are utilized in both, but they differ in content and in *style* of writing.

Three basic rules are common for both technical and business writing. The writer must:

1. Understand the situation and the purpose for writing about it.
2. Identify the audience (recipients).
3. Write for the audience's (recipients') need to know and/or to act.

### 7.4.1 Conciseness

Often, the business writer hides behind verbosity because he or she lacks knowledge of: (1) the topic, (2) the purpose of the letter/report, or (3) the skills in cohesive presentation of the matter. Verbosity should not be confused with extensive detail of information.

Wordiness is not only burdensome to the recipient of the writer's letter, memo, or report, but it also often results in images of phoniness, insincerity, or dishonesty. Wordiness is also the result of overwriting, which occurs when the writer attempts to use what he or she considers higher language. For example, the following phrase illustrates wordiness caused by overwriting: "In the considered opinion of Dr. Smith..." This should be rewritten as follows: "It is the opinion of Dr. Smith... Dr. Smith is a leading authority on the efficacy of...."

Redundancy also results in wordiness (using words unnecessarily). An example of redundancy is found in the following sentence: "The market descended downward."

The pace of writing is important. Each sentence should contain only as much information as the recipient can absorb. Errors in pacing often occur when the writer has expertise in the topic of the letter, memo, or report and does not visualize that the recipient may lack similar expertise, or may not even be familiar with the subject to any degree. Scientists, technical, or financial professionals often transgress this important rule of "pacing."

Direct statements should be used instead of meandering, devious statements. Direct statements are clearer and progress quickly to the writer's objectives. Generally, such direct statements are brief.

Knowing the educational, cultural, and technical experience of the recipient not only provides the writer with the information needed to accomplish his/her purpose, but also determines the language that the recipient can best understand. The language and terms used by the writer must be familiar to the recipient. The writer must first determine if the recipient shares his/her field of language. If so, technical terms and jargon can be used. If they don't share the same field of knowledge, the writer should:

- Explain unfamiliar terms and concepts.
- Define terms and processes in clear language.
- Use analogy to relate the unfamiliar to the familiar.
- Use illustrations as necessary to simplify.

In other words, if the writer and the recipient don't "speak the same language," the writer has to write in a language understandable to the recipient by explaining terms and concepts. There is no common language of the workplace except clear English. It is the writer's responsibility to make the writing communicate, not the recipient's task to translate it.

However, the above needs to be counterbalanced by the guidance provided in the previous chapters of this book, particularly that not everyone needs to understand every aspect of the written communication. In each piece of written communication that is sent to several recipients, some of the individuals will find it more relevant to them than will others.

Verbiage and overwriting may be related. One cause of this is the lack of use of personal pronouns, a practice disfavored by the traditional writing conventions. Verbiage is found in the following sentence: "It was for the CEO's office that I conducted an analysis and evaluation with regard to the feasibility of relocating the Ferrite Core production from Mexico to China." This revised sentence reads better: "I analyzed and evaluated for the CEO's office the feasibility of relocating the Ferrite Core production from Mexico to China."

## 7.5 Tone

An advantage of the business writer is that the recipients must look at least cursorily at the written communication because it pertains to them. Of course, this is contingent in large part upon whether the writer is careful in his/her selection of audience. Therefore, the business writer usually doesn't have to worry that if he/she doesn't use the right "pitch," the recipient will toss the message into the wastebasket. The business writer should bear in mind, however, that there is not always a "captive audience."

One of the writer's first considerations about the audience is the person's relative position to the writer in the organization. Tone is referred to here. The writer's tone should never be condescending or obsequious, but its level of authority can vary according to the position of the recipient.

The recipients of the correspondence possess different characteristics that will affect the manner in which they receive the message. If the writer fails to consider this factor, the message will be ineffective.

### 7.5.1 Bad Tone

Bad tone is the most subtle way to reflect negatively upon the writer. Bad tone, when caused by carelessly chosen words (such as "allege" or "appointment"), is insidious.

Tone is the result of the writer's view of himself/herself in relation to the recipient. Bad tone in correspondence occurs when the writer shifts his/her feelings towards the recipient from neutral, middle ground towards extreme, "personal" feelings.

It is important to make a distinction here. The letter, memo, or report contains *bad tone* only if the words, expressions, or images were carelessly selected by the writer. In contrast, if the writer deliberately selected his/her language, however personalized, then the term "bad tone" does not apply.

Typical causes of bad tone in correspondence are:

- Feelings of inferiority or subservience. This can, and often does, result in unfavorable, even dire consequences to the writer.
- The lesser problem occurs when the writer fawns over the recipient, or uses effusive language in writing to or about the recipient. Examples of this find the recipient in the role of a famous scientist, economist, or public person.

  In fact, very few businesspersons have the occasion to write to or about such a level of "famous" person. However, there are plenty of "home-grown stars" within the corporation itself, in the industry, or among outside professionals about whom business writers might write fawningly or effusively.

  A good example would be the company's current star dynamic General Manager of a very successful subsidiary, particularly in a foreign country, especially if he or she recently came from a "famous" Fortune 100 multinational corporation. Under these circumstances, the writer's image is likely to merely suffer only in the recipient's view.
- A serious problem arises if the writer lets the content of the letter and more so of the internal memo or report become contaminated with feelings of inferiority or subservience. This occurs frequently, especially in internal memos and reports.
- The writer should always avoid using an accusatory tone. Usually, phrases such as, "You claim," or "You allege" create an accusatory tone in correspondence.

> The writer also should avoid intimidation, either by him/her or by the company. Aside from being in "bad tone," this could even result in legal or regulatory actions against the company or even against the writer. If the correspondence is "internal," it results in intracompany or intraorganizational conflicts.

Using the above example, if the writer presents to the recipients of the memo or report the performance of the General Manager in language tainted with the writer's unconscious feelings of inferiority vis-à-vis the General Manager, then the other recipients are likely to obtain a distorted understanding of the true situation. They may even formulate incorrect decisions which they would not have, but for the writer's "bad tone."

In this book it is continuously emphasized from various angles that the business writer must think in terms of the totality of the "business environment." The writer writes to living persons. It is interpersonal communication, in which feelings, egos, assumptions, psychology, reputations, mind-sets, and other facts are integral parts of the communication.

As is shown here, bad tone often is the cause for biased, distorted, and inaccurate messages, which result in incorrect understanding and harmful decisions by the recipients—instead of just in hurt or hostile feelings towards the writer or his/her company (which in itself is harmful). The examples selected here demonstrate the results of "bad tone" arising in real-life types of important business situations, instead of in the simplistic examples of Mrs. Jones being late with her payments, or the company shipping damaged merchandise.

### 7.5.2 Subordinate Tone of Writing

Almost invariably, books on business writing give the business writer the advice to avoid using a subordinate or inferior tone, and not to express the inferiority of his/her position vis-à-vis the recipient of the letter, memo, or report, because this can result in an uncomfortable, demeaning, or unfavorable attitude towards the writer by the reader. This is used as an example of "bad tone" in these books.

However, an overlooked aspect of the business writer's subordinate or inferior mind-set and "bad tone" is its harmful results in business, which are far beyond the mere psychological effects.

With regard to tone, the writer should set aside the position levels and take a position of equality, but one separated by proper distance. In other words, the business writer should take a neutral approach to the message, take care of decorum, and recognize the recipient's higher position by the salutation, and other such reference where applicable, but the writer must not slant the message.

The following examples illustrate the harmful results caused by the business writer being influenced by writing from a subordinate mind-set.

- The Corporate Quality Control Director writes the monthly report on how the various division and subsidiaries met, excelled, or were below planned targets of Q.C. standards. The Director is a staff member and reports at two levels below the Corporate Vice President, Manufacturing.

  The report is addressed to this V.P., with the Corporate Controller, several general managers, and the Profit Planning Director as other recipients.

  The Quality Control Director, conscious of his/her subordinate position, recognizes that substandard Q.C. performance is the result of manufacturing, and writes the report by projecting better than actual performance in the factories. To achieve this, the Director can report correct and accurate facts and data—the favorable slanted interpretations and cause analyses can make the Corporate Vice President, Manufacturing look better than if the report would have been written in an objective, neutral "tone."

They, including this Vice President, rely on the report and draw erroneous conclusions, without taking corrective actions—all the result of this business writer's "bad tone" message.

- The General Manager of the company's Mexican subsidiary knows that the company's President favors the relocation feasibility study of the rectifier production from Singapore to Mexico.

The General Manager is requested to submit to the Financial Vice President a report with his/her evaluation of the various aspects of this relocation to Mexico. Feeling in a subordinate position to the President, the General Manager writes the report and understates the potential foreign exchange exposures and labor unrest probabilities, as well as neglecting to alert the Financial Vice President to probable higher start-up and import costs than in the corporate prepared relocation feasibility study under the President's auspices.

Assume that both of these reports were written in perfect English, with decorum, and no insulting, flattering, subordinate language. Nevertheless, the messages themselves were subordinate in content—in "bad tone."

The business writer should avoid writing in any of the other "bad tones," such as:

- Insulting the recipient or anyone about whom the writer writes.
- Demeaningly writing, such as about the expertise, education, knowledge, or performance of anyone. Critical writing is proper but demeaning writing is not.
- Negative connotations: the writer provides a proper message but adds his/her qualitative negative judgment or allusions, which are "extras" to the message. For example, instead of: "The yield in semiconductor production in Singapore was seven percent below standards this month." The writer includes a connotative negative judgment by writing: "The yield in semiconductor production by the Chinese workers in Singapore was seven percent below standards this month." (In contrast, while "by the Chinese workers" is still a connotative judgment, it would be a positive one if the results were "seven percent above standards.")

  In many cases, negative connotations also navigate the writer into the area of discriminatory interpretations of his/her message, such as in the above sentence.

## 7.6 Decorum of Forms of Address and Writing

Decorum implies proper formality in rules of conduct suitable to the circumstances. Thus, decorum of addressing and writing to a person implies the proper formality in rules or social and business customs in this area.

Decorum is more encompassing than forms of address; decorum has dignity attached to its meaning. Further, if the business writer does not correctly address certain dignitaries or titled persons, it can cause embarrassment to the writer and result in an unfavorable reaction toward his/her company. This should not be dismissed lightly by Americans' customarily assumed position of equality and informality.

The correct forms of address are covered fairly well in such etiquette books as that of Amy Vanderbilt, who herself was for many years in the diplomatic service, dealing with government officials and also with businesspersons.

Exhibit 7.5.1 lists the forms of address to some persons the business writer may have occasion to write to under various business situations. Full street, city, country addresses are always required, but are not shown in the exhibit. However, it is important to show the correct and full name of the country, such as Republic of Korea (for South Korea), People's Republic of China (for Mainland China), and Republic of China, Taiwan (for Taiwan).

A common rule for the business writer when addressing American public officials, ranging from mayor to federal cabinet members (such as the Secretary of Commerce), is the use of "The Honorable (full name)." This applies also to members of Congress, ambassadors, and judges (except those of the Supreme Court of the United States, in whose case it would be, for example: "Mr. Justice Kennedy" with the salutation of "Dear Mr. Justice," without his name—which illustrates the exceptions to the rules).

The proper form that should be used when addressing foreign government officials at the level of cabinet minister, ambassador, or member of the legislature (such as Chamber of Deputies—albeit this varies for some countries) is:

His Excellency (or "Her Excellency" if appropriate)
(Full name)
(Full title)

But even here deviations exist. In Spain and in other countries that share the Spanish culture (most of Latin America and Mexico), it is "His Excellency Señor (full name)" or "Her Excellency Señora (full name)."

For foreign public officials below the above rank, the address should be "The Honorable (full name)." However, there are deviations even in this area. The salutation of "Excellency" is always in proper decorum for those foreign officials who are addressed as "His (Her) Excellency."

It is also proper to use "Dear Sir" or "Dear Mr. Minister (without adding his/her name)." For lower than minister level officials "Dear Mr. (title)" may be proper in many cases, but if the writer is unsure "Dear Sir" or "Dear Madam" should be used.

For American and foreign ambassadors, the salutation should be: "Dear Mr. Ambassador" or "Dear Madame Ambassador" (for a female American ambassador it is "Dear Madam Ambassador").

In all foreign countries, in the salutation, "Madame" replaces for female officials the "Mister" or the "Señor" (in Spanish culture countries). "Madam" is used to address American women.

A few other notes here are warranted:

- Only in the United States should "Esq." be added following the name of a lawyer. (England is the only other foreign country where "Esq." is added after the name of a lawyer, as well as after the names of other professional men and men of nobility without title.)

  In foreign countries, lawyers have "Dr." for doctor of law in front of their names, for example: Dr. Wilfred Kohl. But even here there are exceptions. In Italy, "AVV." or "Dott." before the name replaces "Dr." for lawyers.

- In some foreign countries, it is proper decorum to use the abbreviated title of their profession in front of their names, such as "Ing." for engineer, for example, Ing. Georgeo Cacace. The salutation would then be as follows, Dear Ing. Cacace.
- If the person holding public office, say the Minister of Finance, is also a titled aristocrat (e.g., a Count), such a title is omitted both in the address and salutation. (The person takes on a public persona.)

Note here that the decorum of address is very important to the business writer and his/her company, particularly when writing to or about foreign public persons. In foreign countries, such social formalities are viewed as signs of respect towards the addressee and his/her office (and country, in case of a public official, especially with regard to correspondence coming from Americans).

The decorum of writing style also requires that the business writer should write with formal, eloquent, dignified, learned—one might call it "grand"—language. This is particularly the case when writing to position levels of director through minister or ambassador in foreign countries.

The tone should be impersonal, with a distance between the business writer (and his/her company) and the addressee of the letter—but not subservient or fawning. The emphasis should be placed on the subject, not on the recipient.

Long and involved sentences are proper in such writing, though the writer must avoid fused sentences. The diction should be less concrete than in other forms of writing.

Learned language includes the use of more highbrow (as opposed to common, popular) words, such as "terminate" (instead of "end"), "concur" (instead of "agree"), "facilitate" (instead of "make easy"), "veracity" (instead of "truth"). Foreign phrases, such as "ipso facto," "hors commerce" "ad hoc," and "status quo" should be used, but the writer must ensure that such phrases are used properly.

## Exhibit 7.6.1

## DECORUM OF SALUTATION (ADDRESSING) OF GOVERNMENT OFFICIALS

| TITLE-HOLDERS | SALUTATION | COMPLIMENTARY CLOSING |
|---|---|---|
| **United States:** | | |
| Secretary of Commerce | Dear Mr. Secretary | Sincerely yours, |
| Secretary of Labor | Dear Madam Secretary | " " |
| Under-Secretary of State | Dear Mr. Under-Secretary | " " |
| Assistant Secretary of Transportation | Dear Mr. Smith | " " |
| U.S. Senator | Dear Senator Brown | " " |
| U.S. Representative | Dear Ms. Jones | " " |
| American Ambassador | Dear Madam Ambassador | " " |
| American Charge d'Affairs | Dear Mr. Taylor | " " |
| Governor | Dear Governor Collins | " " |
| State Senator | Dear Mr. Stern | " " |
| Judge (man) | Dear Mr. Justice | " " |
| (woman) | Dear Judge Browing | " " |
| Foreign Ambassador | Excellency | " " |
| | Dear Mr. Ambassador | " " |
| Foreign Minister of Embassy | Excellency | " " |
| | Dear Mr. Minister | " " |
| **Foreign Countries Other Than Great Britain:** | | |
| Minister of Finance | Dear Mr. Minister | Respectfully yours, |
| | Excellency | " " |
| Director of Finance | Dear Mr. Director | " " |
| | Dear Mr. Liang | " " |
| Vice Minister of Finance | Dear Madame Minister | " " |
| Member of the Parliament | Dear Mr. Schmidt | " " |
| Judge | Your Honor | " " |
| | Dear Judge Gaston | " " |
| Count | Dear Count | Sincerely yours, |
| Countess | Dear Countess | " " |

# CHAPTER 8

# Business Grammar

## 8.1 Overview of Business Grammar

The American business writer should make every effort to comply with the rules of English grammar. Grammatical mistakes seldom result directly in erroneous business decisions, though they can result in the information being wrongly conveyed. When a text contains grammatical errors, the reader is often diverted from focusing on the content due to the presence of such mistakes.

Such errors reflect poorly on the business writer. For this reason, and because grammatical errors also reflect unfavorably on the company when the correspondence is "external," the business writer should strive to acquire grammatical proficiency. Alternatively, the business writer should have a qualified person edit his or her writing.

Prior to beginning our discussion of grammar and sentence structure, it is important to state here briefly what grammar is, what it means, and why it is essential to the business writer.

Grammar includes the rules of a language, in this case English, that governs classes of words and their functions and relations within the sentence. Grammar also sets the rules for the combination and interpretation of these elements.

Some authors even state about the rules of grammar that in truth there are no hard, entrenched rules, because any grammatical construction that has become common in usage becomes proper for writing.

In fact, Generative Grammar is a system of linguistic analysis in which the language is considered to have a finite set of rules. It analyzes the grammatical constructions of the language that are acceptable to be considered as being "grammatical," i.e., grammatically correct.

At a last general note here on this aspect of grammar, no one would seriously question the importance of accountants or engineers learning the rules and formulas of their disciplines, for without such knowledge an accountant or an engineer would be unable to function properly. (In fact, without such knowledge, the individual would have never been able to become an accountant or an engineer in the first place.) Likewise, without a sufficient knowledge of the rules of grammar, a person cannot write properly, irrespective of his/her degree of intelligence or level of education.

The business writer should periodically revisit grammar books, and consider grammar from its many aspects in order to acquire an enhanced professionalism in writing. Fortunately, there are some excellent grammar books that cover both the theoretical and practical applications of grammar.

Here, only a very brief discussion of selective elements of grammar is included. The primary coverage here is of *sentences* from their practical aspects.

## 8.2 Sentence: The Unit of Composition

Sentences traditionally are called the units of composition. This is valid only in the grammatical sense, insofar as each sentence has a subject and a predicate (without these two elements, it is not a sentence), and each sentence stands on its own without being part of another sentence.

Sentences make up the written communication, be it in literature or in business writing. Words seldom provide a meaning when they stand alone; to convey an idea or information, they need to be part of a sentence. Nor do incomplete sentences have meanings in most cases, unless they can be understood as parts of the correspondence.

The business writer should use shorter sentences—that is, 20 words or fewer—rather than longer sentences; they are easier to read and comprehend.

However, some writing warrants the use of longer, more complex sentences. The business writer should be careful when including complex sentences in his/her text, and do his/her best to ensure that the recipient understands the sentence in its totality. Complex sentences often express complicated, interrelated ideas and nuances, and frequently readers do not grasp the full meaning of such sentences.

### 8.2.1 Sentence Structure

The main types of sentences are: simple (basic), compound, complex, and compound-complex.

It is helpful to think about these different types of sentences in the following manner: expanding and/or combining or contracting one type of sentence changes it into another type of sentence. For example, when the writer desires simpler sentences, he/she contracts the compound sentence into two simple sentences. As will be illustrated, compound-complex sentences are the most advanced as far as sentence structure is concerned.

It is noteworthy to remind the business writer that sentences are also classified as declarative, exclamatory, or imperative.

- The Simple (Basic) Sentence

  A basic sentence consists of a subject and a predicate, which may be just a verb or it may need an object to complement it. For example:

  The machine stopped. (This sentence has a subject and a verb, and it is a complete thought.)

The general manager met *the comptroller at the airport.* (The first four words in this sentence contain the subject and the verb; however, it does not represent a complete thought. Further information is necessary—in this case the italicized clause—to complete the thought.)

While most simple sentences are rather short, they may become lengthy and elaborate when the subject and/or the verb are modified.

When rereading his/her text, the writer will sometimes find that too many ideas were incorporated into a sentence. In such a case, the sentence should be broken up into several clear, simple sentences.

- The Compound Sentence

  The compound sentence consists of two or more independent clauses, which are joined either by a coordinating conjunction (*and, but, for, nor, or, so,* or *yet*) or by a semicolon. For example:

  The production was at optimum rate, and sales exceeded plan.

  The compound sentence can be used to create balance between two ideas; however, the writer should make sure that a relation exists between the two clauses. Any two simple sentences can be combined by a coordinating conjunction, but if the sentences are not related in terms of content the reader is left baffled. For example:

  We have more time to file our response, and the Chairman retired.

- The Complex Sentence

  The complex sentence consists of an independent clause and one or more dependent clauses that express subordinate ideas. For example:

  Because the president demanded increased productivity, the production supervisors focused on output.

  Flexibility in writing is an advantage gained from complex sentences. It also provides emphasis, as determined by the writer.

- The Compound-Complex Sentence

  The compound-complex sentence consists of at least one independent clause and at least one dependent clause. For example:

  Because of Jack's relative inexperience with the company, I had expected him to be silent during the meeting; however, he participated by providing insightful comments.

This brief discussion of sentence structure is aimed only to remind the business writer about the diversity and complexity of the components of writing.

### 8.2.2 Sentence Fragments, Comma Splices, and Fused Sentences

Since the essential function of the sentence is to state a complete thought, these three errors are of major concern to the business writer.

The sentence fragment is a part of a sentence—not a complete thought—that is punctuated as if it were a complete sentence. For example: After the meeting ended.

A comma splice occurs when two independent clauses are joined by a comma. For example: The meeting ended, everyone left the conference room.

Comma splices can be corrected either by forming two separate sentences or by inserting a coordinating conjunction between the two independent clauses.

A fused sentence, sometimes called a run-on sentence, occurs when two independent clauses are joined with no punctuation between them. It can be corrected in the same manner as the comma splice.

## 8.3 Combining Sentences to Form Paragraphs

Words form sentences, sentences form paragraphs, and paragraphs form the written communication.

Just as a sentence must have a unified theme, so too must a paragraph. The sentences contained in the paragraph must be related to each other.

Paragraph lengths vary widely in the text from only one sentence to almost a full page. The writer should note that a short paragraph draws more attention to the point made therein.

## 8.4 Other Aspects of Grammar

It is not the intent of this book to merely provide the reader with a listing of grammatical rules. Touched upon here are the elements of grammar most applicable to the area of business writing. Topics such as parallel structure, agreement, tense, parts of speech, voice, and case were not covered in this chapter.

The business writer can obtain reliable guidance on each of these elements only from a competently written English grammar guide. (See *Further Recommended Readings.)*

## 8.5 Conclusion

The area of grammar is so rich, diversified, and complicated that even large business communication texts that cover the topic in over 50 pages do so only superficially, and often merely condense the text found in English grammar books without modifying it to the needs of the business writer.

For this reason, the business writer often disregards the area of grammar, for he/she feels that it does not relate to them, especially since the examples in these books are pedestrian and non-business related.

## 8.6 Further Recommended Reading

*REA's Handbook of English Grammar, Style, and Writing.* Piscataway: Research and Education Association 1996.

REA's *Authoritative Guide* to the

# Top 100

# BUSINESS Schools

**Complete, detailed, and up-to-date profiles of the top 100 business schools in the U.S.A. and abroad.**

*Plus in-depth advice on admissions and financing.*

*Research & Education Association*

*Available at your local bookstore or order directly from us by sending in coupon below.*

RESEARCH & EDUCATION ASSOCIATION
61 Ethel Road W., Piscataway, New Jersey 08854
Phone: (732) 819-8880

VISA MasterCard

Charge Card Number

☐ Payment enclosed
☐ Visa ☐ MasterCard

Expiration Date: ______ / ______
Mo Yr

Please ship **"Top 100 Business Schools"** @ $19.95 plus $4.00 for shipping.

Name ____________________

Address ____________________

City ____________ State ______ Zip ______

# "The ESSENTIALS" of LANGUAGE

Each book in the **LANGUAGE ESSENTIALS** series offers all the essential information of the grammar and vocabulary of the language it covers. They include conjugations, irregular verb forms, and sentence structure, and are designed to help students in preparing for exams and doing homework. The **LANGUAGE ESSENTIALS** are excellent supplements to any class text or course of study.

The **LANGUAGE ESSENTIALS** are complete and concise, with quick access to needed information. They also provide a handy reference source at all times. The **LANGUAGE ESSENTIALS** are prepared with REA's customary concern for high professional quality and student needs.

***Available Titles Include:***

**French** **Italian**

**German** **Spanish**

*If you would like more information about any of these books, complete the coupon below and return it to us or visit your local bookstore.*

RESEARCH & EDUCATION ASSOCIATION
61 Ethel Road W. • Piscataway, New Jersey 08854
Phone: (732) 819-8880

**Please send me more information about your** LANGUAGE **Essentials books**

Name ____________________

Address ____________________

City ____________________ State ______ Zip ____________

REA's

HANDBOOK OF

ENGLISH

Grammar, Style, and Writing

**All the essentials that you need are contained in this simple and practical book**

***Learn quickly and easily all you need to know about***

- **Rules and exceptions in grammar**
- **Spelling and proper punctuation**
- **Common errors in sentence structure**
- **Correct usage from 2,000 examples**
- **Effective writing skills**

***Complete practice exercises with answers follow each chapter***

REA RESEARCH & EDUCATION ASSOCIATION

*Available at your local bookstore or order directly from us by sending in coupon below.*

RESEARCH & EDUCATION ASSOCIATION
61 Ethel Road W., Piscataway, New Jersey 08854
Phone: (732) 819-8880

VISA MasterCard

Charge Card Number

☐ Payment enclosed
☐ Visa ☐ MasterCard

Expiration Date: ______ / ______
Mo Yr

Please ship **"Handbook of English"** @ $19.95 plus $4.00 for shipping.

Name ______________________

Address ______________________

City ______________ State ______ Zip ______